Home and Workshop Guide to
Glues and Adhesives

A Popular Science Book

Home and Workshop Guide to

Glues and Adhesives

by George Daniels

POPULAR SCIENCE

HARPER & ROW

New York, Evanston, San Francisco, London

Manufactured in the United States of America

Caution

This book contains instructions for using materials that can cause personal injury or other damage if handled improperly. To promote safety and successful projects, the book advises the reader to take special precautions. But since hazards vary from product to product, the book routinely advises that the reader also study the manufacturer's instructions. On hazardous materials, cautionary labeling and first-aid instructions are required by law. Common label warnings include "flammable," "vapor harmful," and "use with adequate ventilation."

Since neither the author nor the publisher can be on hand to control the actual handling of the materials and processes in this book, neither of them can guarantee the results. Therefore, each of them expressly disclaims any responsibility for injury to persons or property related to the book's use.

Contents

1 | Glue Lore

CHANCES ARE you're within arm's reach of something glued. It might be a desk, a table, a part of your car, or the chair in which you're sitting. Although we seldom think of it, glue is an important part of our world, and it has been used for more than 3,500 years — probably longer.

Found in the tombs of ancient Egyptian rulers, furniture more than 35 centuries old tells much about glue's early use and durability. Museums hold examples of this furniture, surfaced with fine veneers and decorated with ivory inlays — all glued to wooden bases. And such work was not confined to Egypt. There is evidence that glue was used by artisans of many early civilizations — including the Assyrians, Babylonians, Greeks, and Romans. Possibly, the wide use of glue in ancient times and during the intervening centuries stemmed from the variety of natural ingredients from which it could be made. These ingredients included blood, bones, hides, the offal of fish, vegetables, grains, eggs, and milk.

Some of these ancient glues have prevailed into more recent times, even to the present, as will be covered further in Chapter 2. For example, glue derived from bones and hides became the standby of cabinetmakers long ago, and it was used by some of the greatest of them, including Sheraton, Hepplewhite, Chippendale, and Duncan Phyfe. These glues are still widely used today.

But the use of glue was by no means limited to peaceful applications. It has long been used for fletching arrows (attaching the feathers). And though rarely mentioned, its role in more modern weaponry has been great. The laminated wooden propellors of World War I fighting planes were glued together, usually with casein glue, made from milk. The same glue also held the reinforcing gussets to the wing ribs of many of these planes. It was chosen both for its strength and its moisture resistance. These qualities had earlier been proven in the covered bridges the glue reinforced. Here glued framework of the bridges showed no weakening or deterioration after more than 25 years.

But casein was only one of the types that found a place in the aircraft of the day. Inside the giant Zeppelins one of the early flexible adhesives, somewhat akin to our modern cellulose types, held a shingling of gold-

1

Wooden airplane propellors are laminated to minimize distortion. In the early days, casein glue was used to bond the layers together. Today, resorcinol glue is used because it is stronger than the wood itself. The favored wood is yellow birch, as shown here.

This is one of the first two-planked wooden canoes made by the author with brush-applied resorcinol-resin glue and no metal fastenings. Built in the 1950s, the canoe has never leaked or required caulking. It was still in use as this book went to press.

beater skins (animal membranes) to the fabric backing of the hydrogen gas cells.

Yet flexible adhesives derived from cellulose had been used long before the days of the Zeppelins. The most common was collodion, a solution of nitrocellulose in ether and alcohol. In the 19th century printing trade, collodion was often called "liquid cuticle," owing to the way it was used by typesetters. To spare their fingers from the abrasion caused when handling metal type, the typesetters swabbed their fingertips with collodion, which quickly dried to form a tough, flexible protective coating on the skin. And, since collodion was an effective adhesive, it stayed in place.

Chemists, too, saw many possibilities for collodion because it was a form of nitrocellulose, and nitrocellulose could be greatly varied. In highly nitrated form it became explosive guncotton, which Alfred Nobel combined with nitroglycerine to make a smokeless powder far more powerful than dynamite. In the less highly nitrated form called pyroxylin, it dissolved in ether and alcohol to become the collodion that protected the typesetters' fingers. Though chemists had tried and failed, an American typesetter named John Hyatt was first to make collodion into a plastic. By combining it with camphor he produced celluloid and began manufacturing it in 1872. But the story didn't end with the plastic. Celluloid, dissolved in acetone or banana oil, served as an early form of quick-setting cellulose cement, especially useful to model makers. (Banana oil, named for its banana-like aroma, is technically known as amyl acetate. This oil, made by the action of acetic acid on amyl alcohol, is also called fusel oil, an unwanted, poisonous ingredient often found in bootleg liquor of the Prohibition Era.)

In 1910, a factory-made nitrocellulose cement was formulated to hold repair patches on wood and canvas canoes. Colored with an amber tint for easy visibility during application, it was tradenamed Ambroid. This is still sold in hardware stores, along with an assortment of cellulose cements developed later. Though still used for canoe patching and a wide variety of home repairs, Ambroid is probably best remembered for the role it played in the heyday of model airplane building. So widespread was its use that its name popularly became a verb as well as a noun. You "ambroided" parts together.

An avalanche of technological breakthroughs in the 1940s and 50s resulted in new glues whose capabilities old-time glue makers found astonishing. One was resorcinol resin glue. Developed to bond the keels of World War II torpedo boats, it was the first glue so absolutely waterproof that joints made with it could withstand days of boiling without separating. Yet, before it hardened, it could be washed from hands and tools with plain water. In the early stages of development, one of the formulas was known simply as BC 17613. Today, its descendents are in hardware stores under tradenames such as Weldwood Resorcinol Resin Glue.

Another, altogether different form of waterproof adhesive that made its debut during the same period was acrylonitrile Pliobond, originally used to speed the production of tracks for armored tanks. (Previously, steel used for tracks had been copper plated so that rubber would adhere to it satisfactorily.) One of the first stick-to-anything adhesives, this Pliobond had ample strength to handle the tremendous stresses imposed on tank tracks. Yet when set, it was completely flexible — so flexible that it was also used to hold pockets in overalls. It, too, is available in most hardware stores.

Another stick-to-anything adhesive was first made from ground-up tire tubes. Though the formula has been modified since the first production, this one, when set, was stiff but not brittle. Tradenamed Miracle Black Magic, it was used to seal cofferdams to the outside of military cargo ship hulls so welders could handle repairs below the waterline. Now sold in most hardware stores, it still does problem jobs, though usually of a smaller and less glamorous nature, such as replacing loose tiles on bathroom walls.

Some of the World War II wonder glues, however, were late reaching the retail outlets because of the wartime demand. The epoxies, for example, were known to European chemists in the 1930s but didn't reach the market until the late 40s. And some of the post-war adhesives had strange origins. P.A.C. (acrylic-vinyl combination), one of the first super-fast-setting adhesives, evolved from a formula used for dental fillings. Hence it had to cure rapidly, and be completely waterproof. Hardening time is as little as 5 minutes (now matched by some epoxies). Its strength is 3 tons per square inch. And it holds as well in poorly fitted joints as in tight ones. Chapter 2 tells you where to get it.

The fastest-setting adhesives, first used industrially in the 1950s, are the cyanoacrylates, originally developed for industrial production-line use. They were the only type that could set so fast that they hardened before the production line moved past the worker. Now sold in retail outlets under trade names such as Zip-Grip, they actually set in seconds.

Not all of the new adhesives were originally intended to be adhesives; some were intended for other purposes. The polysulfides (thiokols) for example, began as flexible waterproof caulking materials. But since they

This modern version of an all-resorcinol-glued canoe was assembled in a day. Sides are cedar clapboard with 1/2-inch pine gunwale rub strips. The bottom is 1/4-inch exterior plywood. Framing is 1/2-inch pine. Weight: about 70 pounds.

form a high-strength bond with many materials and cure to a form of synthetic rubber, they can also serve as exceptionally resilient adhesives.

The adhesives described so far are just a few in the range now in use. Adhesives play a major role in our lives, some in life support itself, such as the epoxies sometimes used to seal heart pacers. And more adhesives are on the way.

IS IT A GLUE OR AN ADHESIVE? One problem that remains is terminology. Purists still debate as to which bonding materials should be called glues and which should be called adhesives. Some authorities contend that the word *glue* should be reserved for material with an animal base, such as hide glue. Yet the chemically based resorcinol types usually include the term *glue* in their labels. The same is true of the polyvinyl acetate types, such as Scotch Wood & Paper Glue, commonly called white glue. The aliphatic resin types also include the word *glue* in their labels, as in the case of Titebond Glue. Other makers play it safe by combining several terms in one tradename, such as P.A.C., which is the abbreviation for Plastic Adhesive Cement.

Note: In this book we'll use the terms *glue* and *adhesive* somewhat interchangeably—often going with the term used by the manufacturers of best-known brands.

2 | Selecting the Right Glues and Adhesives

TODAY YOU CAN BUY ADHESIVES for practically every type of fastening job and practically every material or combination of materials. There are adhesives that fill the gaps in poorly fitted joints, and others so waterproof that even boiling can't weaken their grip. Some set in seconds to let you hold hard-to-clamp parts together while the glue hardens. Others set over a period of hours to allow time for major assembly work. A few can even be mixed in different proportions to provide different hardening times. There's even one that changes color to signal when hardening has begun. You need only select the right one for the job at hand, using the information in the following pages. For some types of work you'll have a choice of several adhesives, in which case economy or ease of application may be the deciding factors. As in buying meat, choose the least expensive type that meets the requirements. After all, you don't buy filet mignon to make beef stew. And you don't need a high-priced waterproof glue for a living-room picture frame.

As to ease of application, the adhesives you'll be buying come in three general forms. The ready-to-use types, of course, are easiest to work with, if they have the qualities needed for the job. You simply apply the adhesive as it comes.

For some work, your best choice may be a water-mixed form. For those you mix the powdered glue with water according to instructions in fixed proportions.

The completely waterproof adhesives are often two-part types sold in two-container units. You mix the two components in specified proportions just before use. The mix proportions of some of these can be varied to produce different hardening times. The temperature of the working area for most of these two-part adhesives must be at least 70°F. Check the manufacturer's instructions on this point. It's important.

Most of the adhesives described in this chapter are available at hardware stores. But a few of the special purpose types may have to be ordered direct from the manufacturer. The names and addresses of manufacturers (at least one for each type) are included with the adhesive specifications that follow. There you'll find general guidelines for the

6

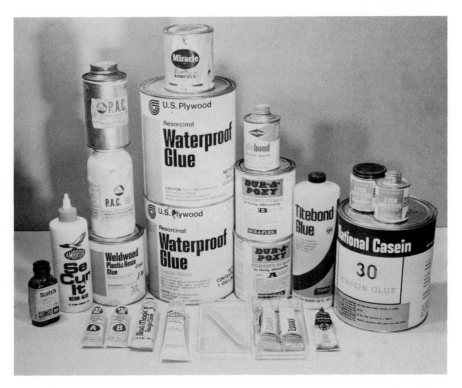

Here are a few of today's popular glue types. Some are ready to use. Some require the mixing of two components. Others must be mixed with water. Most can be bought in hardware stores.

mixing and use of the adhesives. But be sure to follow instructions that come with the brand you buy. Sometimes there are variations in formula between brands, even between different versions of the same adhesive made by the same manufacturer.

READY-TO-USE TYPES

Acrylonitrile base adhesive. This is a creamy tan liquid form with a stretchy-sticky consistency somewhat similar to rubber cement, but with qualities far different from rubber cement. You may have encountered it under the name Pliobond, developed by the Goodyear Tire & Rubber Company. Available in hardware stores and other outlets, it's marketed by the W. J. Ruscoe Company, 483 Kenmore Blvd., Akron, OH 44301. It retails in small bottles, cans, and tubes, though larger containers are available on order.

Acrylonitrile will bond practically any materials by one of three methods. Yet it remains completely flexible and waterproof. (Although it bonds to wood it is not recommended as a wood glue, but may be used to attach other materials to wood.) During the initial drying period it has a slight odor that dissipates quickly. Because its methyl-ethyl-ketone solvent is flammable, the adhesive should be used with fire-safe precautions. The bonding method used depends on the material being bonded and the details of the individual job, as follows:

• *Method 1.* Apply the adhesive to both joining surfaces, allow it to become tacky, and join the parts under enough pressure to maintain good contact. For typical fabric repairs, as in patching work clothes, a sail, or a convertible auto top, you can often press the adhesive-coated surfaces together by hand. This method (wet bonding) is suited to joining porous or semiporous materials that provide a means of escape for the solvent.

• *Method 2.* Coat both surfaces to be joined and allow them to dry completely. Then reactivate one surface by wetting it with methyl ethyl ketone, acetone, or ethyl acetate. Then join the parts with hand pressure. This method (reactivation) is suited to joining larger areas of porous materials. The reactivation step can usually be carried out more rapidly than the original adhesive coating step, and can usually be completed before any part of the coating has begun to dry.

• *Method 3.* Coat both surfaces to be joined and allow them to dry completely, as in Method 2. Then place the coated surfaces together under pressure, and heat at a temperature of from 175° to 325°F. Since most of the solvent evaporates before the parts are joined, this is the method for nonporous materials. It can also be used on porous or semiporous materials to speed completion of the project. On most work, the heat can be applied by pressing a heat-adjustable laundry iron against the back of one of the parts. The heat is then conducted to the adhesive. (Of course, don't use this method if the heat would damage the part you want to join.)

Aliphatic resin glue. This glue is liquid that resembles heavy cream in color and consistency. It dries translucent with a slight honey tone, and can be precolored with water-soluble dyes to approximate the tone of the wood on which it will be used. It's also nontoxic, nonflammable, and nonstaining.

Though now used by do-it-yourselfers, aliphatic resin glue was originally developed for the furniture industry. Like the traditional "hot" animal glues that required a heated glue pot to keep them at usable consistency, aliphatic glue has high "tack" (instant stickiness) but does not need to be heated. In woodworking projects, this lets you press reinforcements in place (such as concealed glue blocks in drawers) without clamps or other fastenings. Because of the tackiness, the block stays put. But try it first on scrap wood because there can be variations from brand to brand. The glue also has a high degree of moisture resistance, though it's not actually waterproof.

Where clamps are required, they can often be removed in as little as 45 minutes because of the glue's rapid setting time. But keep the work evenly supported overnight before putting it into use. Once set, the glue has high resistance to heat and solvents such as lacquers, varnishes, and sealers — minimizing finishing problems. As the glue sets hard, it sands without gumming and clogging sandpaper. It's a good all-round wood-working and cabinetwork glue — with special advantages in poorly heated workshops because it can be used at temperatures from 45° to 110°F. A high-strength bond can be obtained at low temperature, but longer setting times are needed.

Aliphatic resin glue is made by many manufacturers. An example brand is Titebond Glue, made by the Franklin Glue Company, 2020 Bruck St., Columbus, OH 43207

Cellulose nitrate cement. Also called nitrocellulose cement, this has a honey consistency. It is retailed mainly in tubes, although it's available in cans in some brands. Common tradenames are Ever Fast Liquid Cement (clear) and Ambroid (amber colored). The amber-colored cement is useful in two-coat applications, because it shows where the initial coat has been applied. *Caution:* Nitrocellulose cements are flammable.

Application method. On wood where maximum strength is essential, coat both surfaces to be joined. Allow them to dry at least partially before recoating one of the surfaces and pressing the parts together. This procedure allows the cement to penetrate the pores of the wood for a good "bite" and replaces absorbed cement with the final one-surface coat. In most light work, clamping isn't necessary, but parts should be kept in good contact and in alignment until the cement has set. Where the solvent can evaporate freely, typical cements in this category reach 25 percent of their strength in about two hours — full strength in 24 hours. The holding power is as high as 3,500 pounds per square inch — enough for furniture joints that are small enough for application of this fast-drying cement. The cement will bond a wide variety of porous and nonporous materials. Where the cement is used to bond nonporous materials, the solvent escapes through the edges at the glue line. If a large bonding area is involved, however, considerable time may be required

QUICK GUIDE TO GLUES & ADHESIVES

GENERIC & SAMPLE BRAND	TYPE	MATERIALS BONDED	RESISTANCE TO WATER°
Acrylic (P.A.C.)	2-part, liquid & powder	Most porous & nonporous. Make test. See p. 20.	Waterproof
Acrylonitrile (Pliobond)	1-part liquid	Most porous & nonporous. Flexible. See p. 7.	Waterproof
Aliphatic (Titebond)	1-part liquid	Mainly wood & wood products. See p. 8.	Not waterproof
Casein (National Casein Co. No. 30.)	Water-mixed powder	Mainly wood & wood products. See p. 17.	Moisture resistant
Cellulose nitrate (Ambroid)	1-part liquid	Wide range of materials. See p. 9.	Water-resistant
Contact cement (Weldwood Contact Cement)	1-part liquid	Mainly for laminates & veneers to wood or wood products. See p. 12.	Water-resistant
Cyanoacrylate (Zip-Grip)	1-part liquid	Wide range of materials. See p. 13.	Water-resistant
Epoxy (Evercoat Epoxy Glue)	2 parts, both liquid	Wide range of materials. See p. 21.	Waterproof or water-resistant. Fast-set usually has less water resistance.
Hide glue, flake (Cabinetmakers' suppliers)	Water-mixed flakes. See p. 17.	Mainly wood & wood products. See p. 17.	Not waterproof
Hide glue, liquid (Franklin Liquid Hide Glue)	1-part liquid	Mainly wood & wood products. See p. 14.	Not waterproof
Hot-melt glue (Hot-Grip)	1-part solid, liquified by heating	Wide range of materials. See p. 18.	Usually waterproof
Latex combination (Patch-Stix)	1-part liquid	Wide range of materials. See p. 14.	Usually waterproof
Neoprene base (Weldwood Panel & Construction Adhesive)	1-part viscous	Mainly plywood, paneling, etc. See p. 14.	Water-resistant
Polyester (Fibre Glass Evercoat	2-parts, both liquid. Resin and hardener.	Used mainly on wood, bonding fiberglass. See p. 23.	Waterproof

QUICK GUIDE TO GLUES & ADHESIVES

GENERIC & SAMPLE BRAND	TYPE	MATERIALS BONDED	RESISTANCE TO WATER°
Polyethylene hot-melt (Thermogrip)	Solid cartridges	Wide range of materials. See p. 19.	Waterproof
Polysulfide (Exide Polysulfide Caulk)	1 and 2-part systems. See text.	Used as caulk and adhesive on wood and other materials. See p. 15.	Waterproof
Polyvinyl acetate. Abbreviated PVA. (Fas'n-It)	1-part liquid	Mainly wood, wood products, paper, etc. See p. 15.	Not waterproof
Polyvinyl chloride. Abbreviated PVC. (Sheer Magic)	1-part liquid	Wide range of materials. See p. 16.	Water-resistant
Resorcinol resin glue (Weldwood Resorcinol Waterproof Glue)	2-part, liquid & powder	Wood, wood products, plastic laminates, etc. See p. 25.	Waterproof
Rubber base (Black Magic Tough Glue)	1-part viscous	Wide range of materials. See p. 16.	Waterproof
Urea-resin glue (Weldwood Plastic Resin Glue)	Water-mixed powder	Mainly wood and wood products. See. p. 18.	Water-resistant
Water-phase epoxy (Dur-A-Poxy)	2-parts, both liquid	Wide range of materials. See p. 23.	Waterproof

°Note: The glues and adhesives in this table are rated as water-resistant or waterproof in relation to their most common uses.

before solvent escapes from the inner area. You can promote more rapid escape by placing a porous material such as a thin layer of wood or fabric between the nonporous materials. Before using this method on important work, experiment on scrap materials.

The relatively low solids content of these cements results in some shrinkage with drying. This tends to draw the parts of a joint together with a fine, sometimes almost invisible glue line. This is an advantage in many types of woodwork. But when large, thin parts are subjected to this shrinkage, distortion can occur, unless some method of bracing is used during the setting period. (Extra-fast-setting types are also available from model builders' suppliers for quick emergency repairs, as at model airplane field competitions.) Ever Fast Liquid Cement (clear) and Ambroid (amber colored) are from Ambroid Company, Inc., Taunton, MA 02780. Another brand of clear is Duco Cement, by E. I. du Pont de Nemours & Co., Inc., Wilmington, DE 19898.

Contact cement. This type is available in regular and water-base form, retailed in cans and jars. The regular type is a thin, yellowish syrupy form, based on neoprene and other synthetics with a solvent such as toluol or naptho. *Caution:* This type cement is flammable. Water-based contact cement is of similar consistency but it is usually off-white or milky white in color. Because of its water base, this type of contact cement is nonflammable and can be used where flammable types might be inconvenient or dangerous. For example, it can be used where turning off pilot lights of gas equipment would involve technical know-how. Both types of cement are used in gluing methods not common to other adhesives because contact cements are intended chiefly for the application of thin laminates such as sheet-plastic countertops. One smooth-flowing version is also formulated for wood veneer in cabinetwork.

The usual application procedure requires the use of a heavy paper "slip sheet" as described shortly. To start, both surfaces to be joined are coated with the cement and allowed to dry, which usually takes 30 to 40 minutes. When dry, the regular cement has a glossy sheen. Dull spots indicate inadequate coverage or excessive local absorption, and should be recoated. Water-based cements commonly turn from milky white to light tan when dry. Either can be tested for dryness with a piece of paper. If no cement adheres to a piece of paper pressed lightly against the coated surface, drying is complete, and the cement-coated surfaces are ready to be brought together. Although dry, they will bond instantly on contact. So perfect initial alignment is extremely important.

The "slip sheet" of heavy wrapping paper minimizes the chance of alignment error. Place the paper sheet on the up-facing cement-coated surface, covering the entire area. Then place the cement-coated surface of the laminate downward to rest on the paper. The paper will not stick to either one. Align the laminate precisely, as it is to be on the finished job. Then, holding it carefully in alignment, lift the near edge slightly, and have a helper pull the paper outward an inch or two from under that edge. This allows the two cement-coated surfaces to come together along the area exposed by pulling the paper outward, and holds the laminate in position from then on. (Because only a small area is bonded at the initial stage, it is still possible, though not desirable, to peel off the laminate if it should be misaligned.) If all is well, press down firmly on the area of the laminate under which the paper has been removed. Then withdraw the entire paper. Roll the entire laminate surface down firmly with a small, rubber hand roller from your laminate supplier or from a photo supply store to get a thorough bond. Roll from the center of the laminate toward the edges, working out any raised areas that may indicate trapped air. *Do not rush* this type of work by putting the surfaces together before the cement is dry. Trapped solvent will result in loose areas after the piece has been in use for a while. Brands: Weldwood Contact Cement (both types) by Roberts Consolidated Industries, 600 N. Baldwin Park Blvd., City of Industry, CA 91749. For veneer, Constantine's Veneer Glue, by Albert Constantine and Son, 2050 Eastchester Rd., Bronx, NY 10461.

Cyanoacrylate adhesive. This type was discovered by accident in the 1950s during basic research on polymers. In the course of routine measurements, scientists found that the prisms of their measuring instrument were adhered inseparably by the sample. Within hours, the scientists determined that the sample also produced strong bonds on a variety of other materials.

Cyanoacrylic adhesives are available in small tubes. Their outstanding characteristic is the quick setting time, measured in seconds. It bonds metals, as well as rubber and most plastics. It is oil and chemical resistant, but it is usually not recommended for parts subject to great shock or peel. And it is not a gap filler. Yet the extremely fast setting time makes the cyanoacrylates well suited to small area repairs where clamping or otherwise supporting the parts is difficult or impossible. Parts can simply be hand-held in position until the joint is firm. Common brands: Zip-Grip by Devcon Corporation, Danvers, MA 01923, and Permabond, available from Edmund Scientific Co., 1776 Edscorp Building, Barrington, NJ 08007.

Cyanoacrylates usually carry a warning to avoid inadvertent skin bonding, because these adhesives can stick fingers together or to other objects almost instantly. If this happens, powerful pull-apart methods are *not* recommended. Devcon recommends placing the affected parts under running water and working a small rounded object such as a paper clip gently between them to separate them. Bond-Solv available from Edmund Scientific Company offers another means of separating. This is a special solvent formulated to separate parts bonded by certain cyano-acrylates. For separating, the safest bet is to follow the manufacturer's instructions implicitly.

Hide glue (liquid). This is a modern liquid form of the traditional cabinetmaker glue. Unlike the flake forms of hide glue, the liquid form is ready to use as it comes from the container; it requires no heating. Liquid hide glue has a moderately long setting time that allows precise, unhurried assembly of parts without risk of "chilled" joints sometimes encountered with "hot" glues. Heat resistant and unaffected by most lacquers, varnish, and sealers, it's a good, home-shop glue. Example brand: Franklin Liquid Hide Glue, by the Franklin Glue Company, Columbus, OH 43207.

Latex combination adhesives. These take a variety of forms, some completely flexible when set. (In setting, the adhesive actually becomes synthetic rubber, bonding with complete flexibility.) Flexible Patch-Stix is widely used for practically all types of fabric joining and repair. White in color, it can be brushed onto the fabric surfaces to be joined. It is available from Adhesive Products Corporation, 1660 Boone Ave., New York, NY 10460.

Neoprene-base adhesives. Typically, these are used to mount paneling directly to studs or furring strips without nails. They're available in cartridges for use in standard-size caulking guns. On conventional framing, an average cartridge will bond four 4×8-foot panels. Coverage, of course, varies with the method of spreading (continuous bead or intermittent bead) and with the spacing of the furring or studding (see Chapter 8).

The usual method calls for application of the adhesive bead to the framing or furring, then pressing the panel firmly against it. (A few light finishing nails are sometimes used through the top of the panel into the framing. These are only partly driven to serve as a "hinge," permitting the panel to be swung outward from the wall, then swung back without losing alignment.) After initial adhesive contact, the panel is pulled free of the framing to allow the adhesive to air dry for 8 to 10 minutes, increasing "tack." When the panel is again pressed into contact, the bond is strong and permanent. Since there may be some variation between products, always follow the manufacturer's instructions. Example product: Weldwood Panel and Construction Adhesive by Roberts Consolidated Industries, 600 N. Baldwin Park Blvd., City of Industry, CA 91749.

Polysulfide adhesives. These are made in both one-part, ready-to-use form, and in two-part types that must be mixed before use. In the ready-to-use form, the setting time depends on the humidity in the air because the adhesive is moisture-setting. That is, in damp weather it may set in a day or so; in dry weather it may require a week. Since formulas vary, follow the manufacturer's directions and plan on the setting times indicated.

Although the polysulfides are widely regarded merely as seam-caulking materials, many of them (particularly marine types) are also powerful adhesives with very unusual attributes. When set, they are actually synthetic rubber, bonded to the surfaces on which they have been applied. When used as seam caulking in boat work, they bond to the edges of adjacent planks. Here they actually join the planks with a flexible adhesive that stretches to keep the seam sealed when the wood shrinks and compresses when the wood swells and thereby narrows the seams. They can also be used to provide vibration-dampening mountings for workshop motors.

Leftover polysulfide can be formed into rubber washers and other shapes. Molds for this purpose can be made of scrap wood and coated with a suitable parting compound to prevent the molded shape from sticking to the mold.

In seam work, you can prevent the polysulfide from sticking to areas adjacent to the seams by covering those areas with masking tape. It's much easier to peel off the tape than to sand off a layer of rubber. Example brand: Exide Polysulfide Caulk by Atlas Minerals, Mertztown, PA 19539.

Polyvinyl acetate glue (PVA). Often simply called "white glue," this is a polyvinyl-acetate-resin emulsion usually retailed in plastic squeeze bottles of many sizes. Although milk white in liquid state, most brands dry colorless and transparent, forming an almost invisible glue line. PVA is a good glue for cabinetwork but should not be used in joints that are under sustained load where the glue alone must resist the load. (Most furniture joints themselves are designed to provide structural support for sustained loads; the glue merely holds the joint together.) Nor should PVA be subjected to high humidity and high temperature, because both tend to reduce PVA's gripping power. Yet under normal conditions, PVA has high strength and impact resistance.

For application, spread PVA over both surfaces to be joined. Most brands should not be left more than 10 minutes before you join the parts. Corrosion will result if you use PVA on metal or place it in metal containers. Example brands: Weldwood Presto-Set Glue by Roberts Consolidated Industries, 600 N. Baldwin Park Blvd., City of Industry, CA 91749, and Fas'n-It, by the Ambroid Company, Inc., Brockton, MA 02403.

Polyvinyl chloride (PVC). In its usual form, this is a crystal clear, fast drying cement that is unaffected by gasoline, oil, or alcohol. In home repair work it bonds glass, marble, china, porcelain, metal, many hard plastics, and a wide variety of other materials. If you're not sure it will bond a particular plastic try a drop on an inconspicuous part of the material, then wipe the drop off after a brief period. Usually, if the cement has etched the surface, it will also bond the material. Used with fiberglass cloth, it can also stop leaks, such as in plumbing and tanks. The leak area must be dry before the cement is applied. *Caution:* The cement is flammable, so follow manufacturer's instructions.

As with most adhesives, surfaces to be bonded must be clean and dry. Spread the cement evenly with the spreader cap on the tube. Wait 2 or 3 minutes for the spread cement to become tacky before pressing the parts together. No clamping pressure is necessary if the parts will remain in contact and in position. On porous materials, apply the cement to both surfaces to be joined, allowing the cement to become dry to the touch. Then reapply the cement and press the parts together. Where squeeze-out occurs along the glue line, remove excess cement with lacquer thinner. To bond a flexible material, like felt, to a rigid material, like a lamp base, apply the cement to the rigid material only. Example brand: Sheer Magic by Miracle Adhesives Corp., Bellmore, Long Island, NY 11710.

Rubber-base adhesive. Originally made from ground-up tire tubes, this is a "stick-to-anything" adhesive. It bonds to almost any material, including nonporous ones, and forms a firm bond if there is a means of escape for the solvent. It sets hard, but retains enough flexibility to avoid brittleness, and it does not require clamping except when necessary to hold parts in position. Allow 24 hours for the adhesive to develop full strength. Because of its gap-filling attribute and high water resistance, it has been used to seal cofferdams to steel ship hulls to permit welded repairs below the waterline. It is also used to bond signs to masonry walls, reset loose bathroom tiles, and prevent vibration-loosening of nuts and bolts. (Simply spread it over the nut, and work it into bolt threads.) Brand names: Black Magic Tough Glue and Brite Magic Tan Adhesive, both by Miracle Adhesives Corp., Bellmore, Long Island, NY 11710.

Silicone caulk and adhesive. Developed as caulkings, silicones are also adhesives that can take 400 to 600°F. temperatures, depending on type and brand. Check the label. Typical brand: General Electric Silicone Glue & Seal, in small tubes.

WATER-MIXED TYPES

Casein glue. This is made from a protein precipitated from skim milk. In its usual form, the glue is a light-tan powder, usually in cans, that must be mixed with water. Casein glue is an easy-to-use and inexpensive gap-filler type adhesive that can be used at any shop temperature above freezing. Although not waterproof, it is highly water resistant, having retained its bond strength for more than a quarter century in weather-protected joints in the framework of covered bridges. It is also resistant to grease, oil, and gasoline. Casein glue is well suited to most home woodworking projects and is especially good for gluing resinous or oily woods. It is widely used, too, in gluing laminated work, particularly on large jobs such as rafters, where its low cost is important.

Mixing instructions for some brands give the proportions of glue powder and water by volume, typically 1 part glue powder to 1 part water. Others give proportions by weight, typically 1 pound of glue powder to 2 pounds of water. Be sure to add glue to water (not the other way around) and stir rapidly until the powder absorbs the water and becomes thick and pasty. Let this thick mixture stand for 10 to 15 minutes but *do not add water.* During this standing period the water dissolves the powdered glue. Stir again after this, and the thick pasty mix smooths out like heavy cream and is ready to apply. As a general rule clamps should be used for at least the first 4 to 5 hours of drying time. At 70°F, a hardening time of about 8 hours completes the job, though humid air may slow it. If you want to double-check on full hardening, start a test joint of scrap wood at the same time as the work itself, and break-test it after overnight drying. (Test joints are a good precaution in any critical work with any glue.) Example brand: National Casein Company No. 30 Casein Glue by National Casein Company, 601 West 80th St., Chicago, IL 60620.

Hide glue (flake). This is the traditional "hot glue" (readied in heated pots) by cabinetmakers. It is still available by the pound in flake form from cabinetwork supply houses. Although not waterproof, and subject to weakening by dampness, hide glue is well suited to fine cabinetwork not exposed to excessive moisture. It is practically nonstaining and has a shear strength of a ton per square inch. The usual preparation procedure consists of soaking the flakes in water to soften them, then heating with a prescribed amount of water to produce the final hot mix. (Obtain specific instructions for the brand you buy from the dealer.) The glue should be applied to the work while still hot to avoid "chilled" joints that can't be properly assembled because the glue cools and stiffens before assembly. Electric glue pots that heat the glue to the proper temperature and maintain it at that temperature are usually available from glue suppliers. Example brand: Constantine's Cabinet Flake Glue by Albert Constantine and Son, Inc., 2050 Eastchester Rd., Bronx, NY 10461. The same company also stocks thermostatically controlled electric glue pots in pint and quart sizes.

Urea-resin glue. Now widely termed plastic-resin glue, this is commonly sold as a light-tan powder packaged in cans of various sizes for home-shop use. It is prepared for use by mixing with water. Typical proportions by volume are 5 parts resin to 2 parts water. The mixing method usually calls for mixing the glue powder with enough water to make a thick paste, stirring continuously. Then the remainder of the water is added to bring the mix to the consistency of heavy cream, again with continuous stirring. If any small lumps remain, stirring is continued to dissolve them. When smooth, the glue is ready for use.

This is a good cabinetwork glue because it is nonstaining, highly water resistant, and stronger than most wood. Also, it is unaffected by oil, gasoline, and common finish solvents. It is *not* a gap filler type glue, however, and requires well-fitted joints and clamping to assure tight glue lines. (In wide gaps it tends to crystalize and crumble with time. In tightly fitted joints it does not.) As a rule, clamps should remain in place 3 to 6 hours on softwoods, 5 to 7 hours on hardwoods, at around 70°F. At higher temperatures the clamping time can be reduced. On work where appearance is not important, nails may be used in place of clamps. The glue should be spread on both meeting surfaces in a thin coat, and then assembled and clamped while still moist. Example brand: Weldwood Plastic Resin Glue by Roberts Consolidated Industries, 600 N. Baldwin Park Blvd., City of Industry, CA 91749.

HOT-MELT GLUES

Multi-resin hot-melt glue. Like Hot-Grip, this glue is formulated for a variety of gluing applications. It is especially suited to the bonding of plastics, such as polyethylene and polypropylene, that cannot be bonded by ordinary adhesives. The hot-melt glue will bond plastics to each other or to other materials such as vinyl, metal, and wood. To ensure a firm bond, the adhesive contains a combination of resins in a formula that liquifies when heated, makes its bond in the liquid state, then solidifies on cooling.

Hot-Grip resembles butterscotch in color and is packaged in metal pans in which it may be heated for use. Since it hardens by cooling, not by the evaporation of a solvent, it can be used to bond nonporous materials such as glazed tile and steel. *Note:* This works only if you apply the glue and complete the assembly before the glue cools. In all applications the work must be planned, and the parts readied to permit assembly immediately after the glue is applied. If assembly is delayed, the glue may cool and harden before the joint can be made. The glue may also be used to lock splices in polyethylene and polypropylene rope used for towing water skiers and for other marine activities. Example brand: Hot-Grip, by Adhesive Products Corporation, 1660 Boone Ave., New York, NY 10460.

Polyethylene-based hot-melt glue. This is commonly sold in cartridges that must be used with an electric glue gun designed for them. The glue sticks to most materials. It is waterproof, moderately flexible, and very fast setting—forming a bond ready for handling in about a minute. It is light-cream in color before and after application. A budget gun for the glue costs about as much as a good hand saw.

To apply the glue, plug in the gun and allow it to preheat (time may vary with brand). Then insert the glue cartridge, place the point of the gun muzzle on the surface to be glued, and feed the cartridge into the gun with thumb pressure or by trigger action if the gun has a trigger feed. Move the muzzle along the gluing line quickly to permit assembly before the glue cools. Apply the glue to only one of the joining surfaces. Press the other surface of the joint against the glued surface within 20 seconds of application so the glue will not cool before assembly. Hold the parts in position for another 20 seconds, after which they will hold their position. The glue develops 90 percent of its full strength in 60 seconds, the remaining 10 percent in 24 hours. Because of its rapid setting, it is excellent for quick repairs and for work where the parts can be joined progressively, as in gluing gimp to furniture in upholstery work. It is also suited to caulking and seam filling. It should not, of course, be used on jobs requiring application over large areas that might cool before the parts can be assembled. In all work, the glue is simply applied as a bead, just as it comes from the gun. It is *not* spread by spatula or other means. Spreading takes place naturally when the gluing surfaces are pressed together. Manufacturers of both glue and gun: Bostik Division, USM Corporation, Boston St., Middleton, MA 01949, and Swingline Inc., Long Island City, NY 11101.

TWO-PART ADHESIVES

Acrylic resin glue. This is packaged in two parts, one a powder, the other a liquid. It sticks to almost any material and is one of the strongest and fastest setting of all adhesives. Of the generally available types, only hot-melt forms and cyanoacrylates set faster. Acrylic resin glue is also one of the highest-priced adhesives, however. Twin-container packages range in size from 4 ounces to 1 gallon. (Special forms of the resin are available for dental work.)

The glue is exceptional for special marine uses, especially those for which other marine glues are unsuitable. But the rapid (5-minute) setting time generally limits it to jobs that can be coated and assembled quickly. Since the full curing time for some mixes is as long as 30 minutes, these glues can be used for longer periods than many others. The safest procedure is to mix a small test batch and try it for spreadability at 30-second intervals until it becomes unworkable. The manufacturer's instructions will indicate the proportion that provides the longest and shortest working times. One brand, P.A.C., changes color from white to distinct yellow when the mix is no longer workable.

The glue can be used to repair a broken spar when the angle of the break is long enough to make repair feasible. In this case, the broken ends are fitted together dry, then moved apart about $3/16$ inch. After the bottom of the opening is sealed with self-adhering tape, a thin liquid mix is then poured into the gap. The parts are then moved together. No clamping is necessary unless for alignment. This method is designed to coat a relatively large and complex gluing surface in the shortest possible time. Temperature of the workroom may be relatively low because the mix generates its own heat. Before using the method above, check the manufacturer's instructions.

Since the glue has unlimited gap-filling ability, it fills gaps left by split-off fragments. It can even be used to fill large holes. As to strength, at least one large sailing ship has sailed around the world with a spar that had been repaired with acrylic adhesive.

In normal wood gluing the surfaces to be joined are sometimes given a quick prebrushing with the pure liquid component prior to applying the mixed adhesive; this increases penetration. Whatever the material being glued, the meeting surfaces should be thoroughly clean and slightly roughened if possible. Once set, the glue is completely waterproof and unaffected by oil, gas, and many common solvents. Tools and containers can be cleaned with such solvents, however, while the glue is still wet. Hardening is so rapid that disposable containers and applicators are advisable.

To bond metals, the joints may be made with adhesives alone in most cases. Where extra strength is necessary, the joining parts may also be drilled and bolted, or otherwise provided with added mechanical strength. The durability of the adhesive in metal repairs has been demonstrated in repairs of carburetors, gas tanks, and crank cases.

Example brands: 3-Ton Adhesive, whose name indicates strength per square inch, and P.A.C. (Plastic Adhesive Cement) which contains hardening yellow color signal. Both are from Tridox Laboratories, 212 N. 21st St., Philadelphia, PA 19103.

Epoxy glue. This is a two-part glue consisting of a liquid resin and a liquid hardener. Because formulas vary, the parts must be mixed in the proportions specified by the manufacturer. When packaged in tubes, the two parts of the epoxy glue are usually formulated to produce a stiff consistency and to harden properly when mixed in a 1-to-1 proportion. Typical package sizes range from as little as 30 cc in tubes, to gallon cans for marine use. Owing to the honey consistency of the unmixed resin in can sizes, it is often thinned slightly with additives to ease spreading.

If the resin is to be used between nonporous materials, however, it should be "100 percent solids," which means that it consists only of the resin and its hardener, with no solvents or thinners that must evaporate before mating of the surfaces. Then, when mixed, the entire quantity of both liquid components solidifies. Since nothing is lost from the mixture, it can harden without evaporation, and it does not shrink. The usual color of the pure resin ranges from clear to honey, unless colored for mixing purposes.

Surfaces to be joined should be thoroughly cleaned and slightly roughened, if possible. The glue should be applied to both meeting surfaces. Because of the toughness of the hardened glue, squeeze-out should be removed from the work with an acetone-dampened cloth before the glue hardens. This reduces the amount of sanding required later.

Working time and curing time varies with the particular adhesive. In a typical instance, you can figure on a working time of about 2 hours, a tack-free drying time of about 3 hours, and a complete cure in about 18 hours at 70°F. Some formulas differ very widely from these figures. So follow the guides supplied with the one you use. Though very slow-setting epoxies take a week to harden, some brands can be painted or finished as soon as the glue is tack-free. But check labels.

Once set, epoxy is completely waterproof and unaffected by most common solvents and many acids. It can be removed from tools with acetone while still liquid. But since the job is slow, you'll be wiser to use disposable brushes, applicators, and mixing containers.

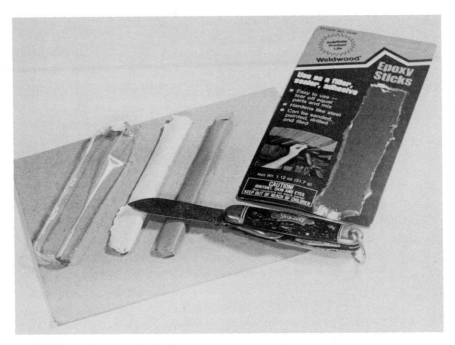

Epoxy in stick form (manufacturer calls them epoxy sticks) can be used as adhesive, filler, or sealer. To use, you cut off equal lengths of the white and the colored sticks (they're soft) and mash them together until color is even, not streaked. Then apply. As adhesive, it works best in a thin layer. For filling or shaping jobs, it can be pressed in place or shaped with a putty knife or spatula. Epoxy sticks are made by Roberts Consolidated Industries, Kalamazoo, MI 49001.

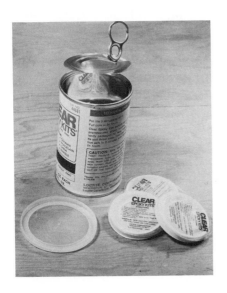

These clear epoxy kits are packaged in rip-top cans (10 mix-it kits to a can). After the can is opened, the snap-on plastic top reseals it. Each wafer-shaped "kit" does a single repair job. Just peel off the plastic film, mix components thoroughly, and apply. The kit is by Loctite Corp., Newington, CT 06111.

You can use epoxy on almost any material, including wood, to provide a permanent bond of unusually high strength. In wood gluing, however, other glues (such as resorcinol, described at the end of this chapter) can provide practically the same or better qualities and strength, often at lower cost. So, unless special problems are involved, they provide sensible economy.

Twin-tube packages from numerous manufacturers are available at hardware stores. Marine suppliers often stock cans of Evercoat Epoxy Glue by Fibre Glass-Evercoat Co., Inc., Cincinnati, OH 45242. For formulas suited to special purposes contact sources such as Miller-Stephenson Chemical Co., Backus Ave., Danbury, CT 06810.

Water-phase epoxy. Though used largely as a coating, this may also be used as an adhesive in numerous applications. Water-phase epoxy is a two-part adhesive, consisting of a liquid resin and a liquid hardener to be mixed according to the manufacturer's instructions usually in equal parts. It differs from other epoxies, in that water washing can remove it from brushes and tools before it hardens. Once hard, the material is completely waterproof, and it can be painted. It is not generally used as a wood-to-wood glue, but is used primarily for such jobs as binding fiberglass over porch or deck seams. In one instance, I bonded fiberglass tape over a seam in painted plywood porch decking with Dur-A-Poxy 200, shortly after a rain shower. The tape was still in place seven years later with no sign of separation. A conventional 100 percent solids epoxy under the name of Dur-A-Glaze is also available. But this one is not a water-phase type. For further information, contact Dur-A-Flex, 100 Meadow St., Hartford, CT 06114.

Polyester resin glue. Often called simply polyester resin, this consists of a liquid resin and a very small amount of liquid catalyst which must be mixed to activate the material just prior to use. After hardening, the resin's color ranges from clear to slightly amber, though it can be pigmented to produce practically any color and eliminate the need for painting in boat fiberglassing.

Working time varies with the amount of catalyst used and with the temperature of the working area, but averages around 30 minutes. So the resin should be mixed only in amounts that can be used up promptly. In fiberglassing work, this usually means mixing batches of less than a quart. To assure accuracy in mixing, the quart-size cans containing the resin and the plastic bottles containing the catalyst are both marked off in fractions so that less than their full content can be mixed. Gallon cans are not usually marked because the resin should be decanted into smaller containers for mixing. Yet catalyst containers for gallon-size resin cans are marked, usually in quarters. For resin amounts smaller than the fractions provided for in the quart-size graduations, you can use disposable graduated medicine measuring cups sold in drugstores. For the catalyst use an eyedropper. To graduate the eyedropper, first fill an empty catalyst bottle with water to the level of one of the graduations. Then draw out water with the eyedropper to reduce the level by the fraction of the bottle graduation desired. Do this at a section of the bottle where the sides are a straight cylinder (not at the rounded upper or lower portions) so that fractional measurements can be made with a small ruler. The eyedropper can then be graduated with a ruler accordingly, using colored nail polish on the glass. The reason for precise measurement is that the amount of catalyst can be critical in small batches. Here many polyester formulas call for as little as a fraction of an ounce of catalyst per gallon of resin. In any event, *follow the directions of the manufacturer.*

A working temperature between 70° and 80°F is best when using polyester. At lower temperatures more catalyst is required, as indicated by a temperature-mix table for the particular resin type. Too much catalyst may weaken the hardened resin. So a "promoter" is used instead beyond a specified catalyst limit. If large areas of the fresh resin coating are exposed, as in boat-hull work, the job should be done in shade rather than direct sunlight. The reason: Sunlight speeds hardening, making estimation of hardening time difficult. To play safe, figure hardening time on the short side, and discard any remaining resin as soon as it shows any tendency to thicken. If the resin hardens on the work before it can be spread, the resulting hardened mass must be removed by sanding — a difficult and time consuming task. Highly humid weather often delays hardening. If you are working under such conditions note the hardening time as you work, and plan accordingly. *Caution:* The resin is flammable in liquid form. Once set, in typical boat lay-ups, it is about as combustible as plywood. Professional boat workers often spread the resin with their bare hands, but you shouldn't use bare hands if you have a history of allergic reaction. And be very careful to protect your eyes from resin spatter, and particularly from the methyl-ethyl-ketone-peroxide (MEK peroxide) catalyst. Indoors, carefully follow the manufacturer's precautions about ventilation.

Polyester resin adhesive will stick to a variety of materials, but it is most widely used in the building of fiberglass boats and in bonding fiberglass and other hull fabrics to wooden hulls. Of course, it is also a highly effective wood glue for work that can be handled within its working time. It is also widely used in stonework repairs, such as in mending broken marble furniture tops. Many brands are available through marine supply dealers and mail order as well as from manufacturers such as Fibre-Glass-Evercoat Co., Inc., Cincinnati, OH 45242.

Resorcinol resin glue. This is packaged in twin cans, one containing the syrupy, burgundy-colored resin, the other the powdered catalyst. The two components are usually mixed in the proportions of 4 liquid to 3 powdered catalyst. Thinning may be done when necessary with a very small amount of alcohol or water. The twin container sizes range from quarter-pint to gallon size at retail outlets.

Before hardening, resorcinol glue can be washed easily from tools and containers, using plain water. After hardening, it is completely waterproof and unaffected by water, gasoline, oil, mild acids, alkalis, or common solvents. Even prolonged boiling does not weaken it. Resorcinol is a leading boat-building glue, stronger than wood at temperatures from minus 40°F to temperatures high enough to set the wood afire. Hardening time varies from 10 hours at 70°F (the minimum safe working temperature) to 3 hours at 100°. It can be sanded and painted as soon as it is hard, though it continues to gain strength for several weeks. It is a moderately good gap-filler glue and can be held together either by clamping or by nails or screws. The fastenings may be left in place or removed after the glue sets. You can judge the hardness of the glue by feeling droplets of squeeze-out. When they "click" off, the bond is strong enough to be put in service. The working life of the mixed glue is about 3 hours at 70°F. But to play safe, plan your work to be completed in a somewhat shorter period. Here, working time can be shortened by temperature variations and sunlight. In hot weather you can extend the working time (or maintain the 70°F figure) by setting the glue container in a bowl of cracked ice. When mixing, be especially careful to avoid inhaling the powdered catalyst or allowing it to get into your eyes, for it is caustic. The powder tends to pack during storage. So before mixing the two components, shake the closed can of powdered catalyst vigorously to "fluff" it. This ensures that the measured proportions will be accurate. *Important:* Use a *separate* measuring implement for the two parts, and be *very careful* not to allow any of the powdered catalyst to get into the can of unused resin. Even a slight amount of the powder in the unused resin can gradually stiffen it and make it unfit for use. Wash all mixing measures thoroughly after use to prevent inadvertent contamination of subsequent batches. The glue is chemically related to that used in waterproof plywood, and is comparable in performance. It was developed originally for use in torpedo-boat construction. Example brand: Weldwood Resorcinol Waterproof Glue by Roberts Consolidated Industries, 600 North Baldwin Park Blvd., City of Industry, CA 91749.

3 | Gluing Tools and How to Use Them

GLUING TOOLS can be classed in three general categories. First, in order of use, are those needed to prepare the adhesive, if preparation is necessary. Next, are the tools used to apply the adhesive. Last, and often extremely important, are the clamping, pressure tools. In some work, too, heating devices are used during the pressure period to speed the setting of the adhesive or to bring about the bonding action. In the case of the hot-melt glues, of course, the heating device is a basic.

It's important to know the requirements of the adhesive you're using and to have all the needed tools on hand before starting the gluing job. As explained later, the tools vary widely according to the adhesives and the working conditions, and may be simple household items such as a pair of matched tin cans or a bowl of cracked ice. For some adhesive work, no tools are required. If you're planning an important job with an adhesive you haven't used before, this chapter will help you get started. But remember that experience is the best teacher. So it's wise to test an unfamiliar adhesive by trying it on a sample of the materials you'll be gluing. Then keep a mental note of each phase of the procedure, including setting time, so you can adapt your test technique to the actual job.

TOOLS FOR MEASURING AND MIXING. If you're using a two-component adhesive or a water-mixed type, careful measuring and thorough mixing are extremely important. In some two-part types, as described later, an excess of one component can cause the adhesive to harden before you have time to apply it. Inadequate mixing, on the other hand, may actually prevent some adhesives from hardening almost entirely. Yet the careless use of measuring tools may render the unused portion of the same adhesive useless by causing it to partially harden in the container. More about this shortly.

MIXING PRECAUTIONS. For two-component adhesives, such as epoxy or resorcinol, use a separate implement to remove the required amount of each component from its container. For small quantity mixes you can use a pair of cheap metal tablespoons or teaspoons. For larger quantities use ladles. The twin implements are definitely essential in order to

Always use separate measuring implements for two-component glues to eliminate the chance of carrying a trace of one component back to the remainder in the other container, eventually spoiling that unused portion. Label or color-code the handles of measuring implements to avoid mixups. Cheap spoons and ladles can give you the usual amounts of glue mix.

avoid contaminating components remaining in the two containers. To be sure you are using the twin implements consistently with the same adhesive component color-code or label the handles. Only a small amount of contamination can result in gradual jelling or stiffening of the leftover adhesive, making it unfit for use. This can be costly because most two-component adhesives are in the higher price range. If one component is a powder and the other a liquid, such as resorcinol, be especially careful not to let the powder air-drift into the liquid. Keep open containers well separated.

MEASURING BY WEIGHT OR VOLUME. The instructions for mixing adhesives may vary. Some tell you to measure the proportions by weight. Others specify proportions by volume. Still others give you the

If powdered glue and water must be proportioned by weight, a postage scale can be used for small amounts. Plastic disposable tumblers, as shown, are lightweight and handy for the purpose, and can be reused. Weigh the tumbler first and subtract its weight from the weight of ingredients. For larger amounts of glue, use disposable paper paint buckets and a kitchen scale.

Keep your glue mixing tools where they're needed. Here, a glass medicine dropper is held to the liquid-component container of an acrylic two-part adhesive. The dropper is essential because the adhesive is usually mixed in small amounts, proportioned according to desired hardening time. The powdered component can be measured with an old kitchen measuring spoon.

option. If you must measure by weight you can use a postal scale for small amounts — a kitchen scale for larger ones. Measuring by volume, of course, is easier. If the instructions are for weight measurement only, you can measure the volume of each component after weighing, and thereafter use the volume method. If one component is a powder, however, it's important to "fluff" it by thoroughly shaking it in its closed container before measuring it. This is because powdered components tend to "pack" in storage. If you measure the powder in the "packed" state, your mix will have an excess of the powdered component. Then after repeating mixings you'll run out of powder before you use up the liquid. Besides, using packed measures, you'll be mixing wrong proportions. The bond-strength effect of the inaccuracy depends on the type of adhesive. After fluffing any powder in its container, allow several minutes for it to settle before opening the container. This way, you'll avoid dust sneezes and also eliminate air-drift of the powder into the other components. Inhaled dust may also be harmful.

If you're measuring a glue such as resorcinol, by volume, you can use glass measuring utensils of the ordinary kitchen variety because the glue can be washed off the utensil with water before the glue hardens. If you're measuring a type that's difficult to wash off, such as epoxy, by all means use disposable measuring utensils. Vegetable or juice cans are a good choice because you can choose a size suited to the amount of adhesive to be mixed. Many juice cans have rings pressed into the metal that can aid in measuring. When specific volumes of a two-part glue such as epoxy must be measured into cans, you can establish the correct level by using a glass measuring cup, wooden strips, and old food cans, as explained under the photo on page 30. The strips should be thick enough to serve also as mixing paddles, after the components are combined in one can. (For P.V.A. glue, use glass mixing bowls, not cans.)

For mixing, use a can or container that provides enough room for stirring without spillage. To do a thorough mixing job with two-component adhesives, you must not only stir the mix but also lift the paddle at fre-

On large gluing jobs with fast-hardening glues, such as resorcinol, you can stretch working time before the glue stiffens by placing the container of mixed glue into a bowl of cracked ice. Always keep a small jar or can close to the glue container so you can park the glue brush in it when necessary.

Juice cans make good glue mixing containers. When mixing two-part glues by volume, the usual method is to use water first to establish correct proportion levels. To do this, stand a wooden strip in the can and pour the measured amount of water into the can. Do the same for both components. Pencil-mark the level for each one on a separate wooden strip, unless components are used in equal amounts. Then dry the can and use new, dry wooden strips marked to match originals. These will show you the correct levels of glue components needed. Since only water is used in the kitchen measuring cup, there are no cleaning worries.

quent intervals to assure that the lower portion of the mix is blended into the upper portion. If you have any doubt that the components have been thoroughly mixed, mix them a little longer. If you're mixing a fairly large batch of adhesive, you can do the job with a paint-mixing accessory designed for power drills. In this type of mixing, however, try to minimize the formation of air bubbles. (*Note:* Epoxies will be difficult to clean from the accessory. Hand-mix flammable adhesives.)

Typically, inadequately mixed two-part adhesives require much more time to harden. Or they may not fully harden at all. Fast-setting two-part adhesives, such as some acrylics and epoxies, must of course be mixed and applied quickly if you are to complete the work before they set. But for the most part, fast setting adhesives are mixed in relatively small quantities, making fast but thorough mixing easy. Usually, too, it's easy to see when components are thoroughly mixed. The fast-setting epoxies are usually different colors that show an even tone when completely blended, usually in a minute or so. The acrylic fast-setters described in Chapter 2 are made up of one powdered and one liquid component, so thorough blending is easy to watch for.

WATER-MIXED GLUES. Casein and urea-formaldehyde (plastic-resin) glues are the water-mixed types most commonly used in the home workshop. Both types are mixed with ordinary tap water, though with casein it's best to have the water near room temperature, or just a little below, but not extremely cold. If the water is 60°F, for example, the mix will warm up somewhat due to chemical action, so your glue will be around 70 to 75°F when thoroughly mixed—just about right for application. Chilled water tends to lengthen the time required for the glue to reach working consistency because casein goes through a sharp consistency change during mixing. When first mixed, it becomes extremely thick and can even stall some electric mixers. At this point the mix appears to need more water, *but do not add water.* Instead, let the mix stand for about 10 minutes and then resume mixing. With this second mixing, the glue turns smooth and creamy and is easy to apply. If you're mixing it with a wooden paddle, leave the paddle in the thick glue so that you can use it for the second mixing. If the creamy glue (after the second mixing) is still too thick for easy spreading, you can add a little water and stir it thoroughly into the mix. Don't add too much water or else you may end up with glue that's too thin.

Urea-formaldehyde glue, a plastic resin, is mixed with tap water too (commonly 5 parts resin by volume, 2 parts water). By weight the mix would be 10 parts powder and 6 parts water. This glue does not require two-stage mixing. If the mix turns out to be too thick for smooth applica-

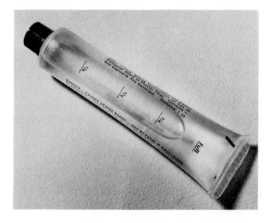

When using polyester resin *be sure* you add the catalyst accurately. Plastic squeeze tubes of catalyst are usually marked in fractions like this, for use with corresponding fractions of resin in a larger container. Too much catalyst can harden resin before you have a chance to use it. Too little may result in delayed hardening, poor strength, or no hardening at all. Follow all cautions on the resin container. For your safety, ventilation of the work area is very important.

tion, you can add a little water. But, as with casein, don't overdo it. You can use a wooden mixing paddle or a drill-powered mixer. For the average job a disposable mixing paddle is best because the glue has a stickiness that requires time-consuming cleaning of tools.

Hide glue (flake type) is also a water-mixed glue, though it's not mixed like casein or urea types. Typically, you mix 1 part flake glue with 1½ parts water, and let this soak for half an hour or more, then place it in a temperature-controlled glue pot set for 150°F, and stir at intervals until smooth. A wooden paddle is the usual tool for mixing. The glue pots made for this purpose are normally water jacketed like a double boiler for smoother temperature control with minimal risk of overheating. One source of the pots is Albert Constantine and Son, Inc., 2050 Eastchester Rd., Bronx, NY 10461.

SURFACE PREPARATION. Before applying all adhesive to surfaces to be joined, be sure you've prepared the surfaces properly. The importance here varies with materials and the adhesives, but you'll usually find necessary measures in the manufacturer's instructions. In all cases, it's vital to have clean, grease-free surfaces. In new work with wood glues the pores of the wood should be opened by light sanding with medium-grade abrasive paper just before the glue is applied (All sanding dust should be removed.) Sanding is especially important where the wood shows any signs of glazing, as result from surfacing with a dull blade. Warp and cup of lumber should be no more than can be straightened by clamping pressure. In repair work, old glue should be sanded away where possible.

APPLICATION AND CLEANUP. The simplest gluing jobs are the small ones in which the adhesive can be applied directly from a squeeze bottle or tube. But for delicate work, such as model building, even the tube tip can be too big for the job. In this case, you can make your own applicator by sharpening a pencil-thin strip of wood and using the point to apply the glue. This method is widely used on model planes and ships.

If you're working on porous materials with cellulose-nitrate cement where peak strength is essential, apply the cement to both of the surfaces to be joined, and allow it to become tacky. Then apply a second coat to one surface and assemble the joint. This method takes into account the portion of the cement absorbed into the surfaces and provides additional cement to bond the two surfaces. Hence it's a method widely used on absorbent materials such as balsa wood. Spills or smears can be cleaned up

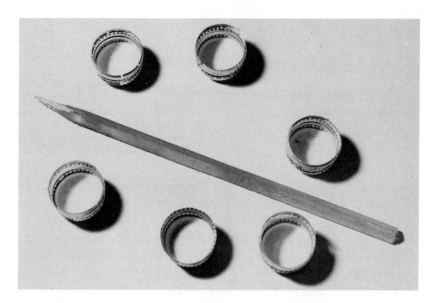

For small mixes of two-component adhesives such as acrylics, you can adjust for hardening time by varying proportions. For this, use screw-on bottle caps. Before mixing adhesive in them, remove the plastic gasket with a rounded-tip knife blade. A slim, pointed wood strip makes a good mixing tool and applicator. Use the point to apply tiny amounts of adhesive. Use the blunt end for larger amounts.

with acetone. But avoid getting the cement on finished surfaces because it can mar them, as can acetone.

Glues, such as resorcinol and casein, can be washed off with water before they harden. So these glues can be applied with reusable brushes. But be sure you wash your brushes before the glue begins to set. Some glues that can be washed off with water before hardening, such as plastic resin glue, still may be so sticky that they require considerable cleaning effort. To decide whether you want to use disposable brushes, mix a small amount of the glue and try washing it off the mixing paddle. Resorcinol is usually easy to wash off before hardening begins. Once hard, it can't even be boiled off.

Epoxies used on small areas, as from twin tubes, are usually applied with a disposable wood strip, which can also be used first for mixing. For large epoxy jobs requiring brushes, cleanup fluids are available with solvent bases such as methyl-ethyl-ketone. You can usually buy these preparations from the maker of the resin. But don't be surprised if the cleanup requires considerable time and effort. For this reason, you may prefer to use disposable brushes. The same is generally true of polyester resin. Most jobs, as in boat work, can be planned to get the maximum use

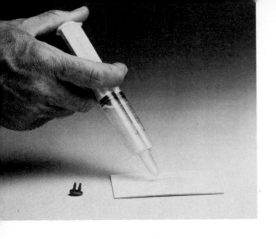

This two-part epoxy in "Dev-Tube" made by Devcon Corporation, emits two components in equal parts when the plunger is pushed. All you need do is mix the parts with a wooden paddle.

from each disposable brush. The twin hulls of a 16-foot catamaran, for example, were fiberglass covered with epoxy that required only four disposable brushes. It's wise, however, to buy a can of the cleanup solvent to remove resin accidentally spilled on tools or smeared on hands. And, of course, follow all manufacturer's safety cautions. Ventilation is especially important if resins are used indoors. Don't assume a resin's vapors are harmless simply because they aren't especially strong.

CLAMPING. Generally, when using glues made primarily for wood, the tighter the joint or the glue line the stronger the glue bond. That's why you may need clamps. Properly used in well-fitted joints, most modern wood glues can form a bond stronger than the wood itself. Yet in the instructions accompanying some wood glues (particularly those used industrially), you may find clamping pressures specified in pounds per square inch, which most of us have no means of measuring. So a popular rule of the thumb calls for clamping the parts together as tightly as possible without damaging the wood. This usually works out well because low-density woods require less clamping pressure than high-density ones. (As a guide, an industrial casein glue specifies 100 to 150 pounds per square inch for low-density woods, 150 to 250 pounds per square inch for high-density ones.) So in gluing most softwoods you can use somewhat less clamping pressure, sparing the softer surface from possible clamp damage. Of course, pads of scrap wood should always be used between metal clamps and the work, so any clamp marks will be on the scrap wood.

On long glue lines, space and tighten your clamps to apply pressure as evenly as possible. The glue "squeeze-out" is a good indication of the evenness of pressure. If the squeeze-out bead is about the same size all the way along, you can conclude that the pressure is evenly applied.

On work where appearance isn't of prime importance, as in many boatbuilding jobs, you can use nails or screws to apply pressure in situations that would otherwise call for a great number of expensive clamps. An adequate number of clamps is an asset to any workshop, yet, some one-time jobs may require more than you'd be likely to need again. That's where nails or screws may do the trick. Where appearance is a prime consideration, finishing nails are sometimes used, driven through small squares of 1/4-inch plywood. After the glue hardens, a pincer type nail puller can be used to bite through the plywood squares and remove the nails. As finishing nails are relatively small in diameter, the nail holes are easily concealed with suitable wood filler.

There are other tricks, too, when you need just a few extra clamps on a one-time basis. Sometimes a monkey wrench can be tightened on the work to serve as a clamp. The same applies to clamp-on kitchen appliances. A clamp-on meat grinder, for example, can often serve as a gluing clamp.

CLAMPS

Called a handscrew, this type of clamp is widely used in woodworking because its jaws are adaptable to various angles. The threaded spindles are simply tightened individually. Wetzler Clamp Co.

This variation of the Wetzler handscrew has aluminum-faced jaws from which most wood glue can be easily removed without scraping.

(Continued)

AND MORE CLAMPS

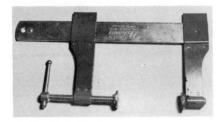

Stanley Tools' bar clamp has a threaded shaft that can be moved to any position along the bar. The clamp locks in place on tightening.

This variation on the bar clamp is called a piling clamp. It's used when a number of duplicate pieces of work are to be glued simultaneously. On this model by Wetzler, the threaded shaft end is fixed. The other end is adjustable along the bar.

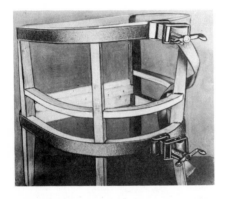

Band clamps employ bands of nylon webbing to clamp irregular work that can't be clamped with jaw-type clamps. Here a Wetzler band clamp is used on a chair.

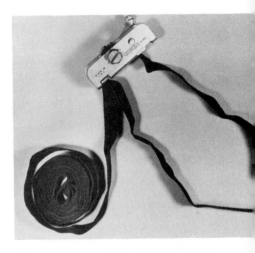

This Stanley web clamp (band clamp) is only a little larger than a cigarette pack when the 12-foot length of webbing is retracted. You can tighten it around work with screwdriver or wrench. A ratchet holds the webbing in place.

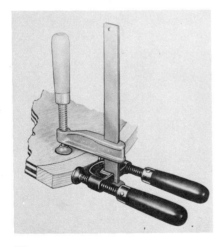

Wetzler cross clamps are used for side gluing, as for the fastening of edging to a tabletop.

STANLEY CORNER CLAMP

The Stanley corner clamp can be used at each corner of a mitered assembly, as shown. This simple setup calls for only two pairs of clamps.

A slot in the corner of each Stanley clamp permits use of sandpaper folded with abrasive side out between mitered parts. For this, set clamp so sandpaper fits snugly. Then slide the sandpaper back and forth to trim the miter to a perfect fit. A backsaw or a wood file may be used in the same way. After trimming, reset the clamp to hold the miter tight when gluing.

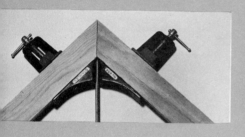

If the mitered assembly is square (not rectangular) and can be worked on a flat surface, you can use a threaded rod and nuts through clamp slots at diagonally opposite corners to hold the assembly together with only two clamps. This method requires care. The rod should not be tightened more than necessary for good contact, or else it will spring the assembly out of square.

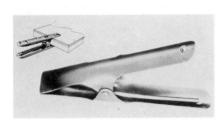

Spring clamps like these are used for quick clamping where relatively light pressure is required, and where fast application and removal are important.

If you plan major cabinetwork on a continuous basis, veneer press frames like this one by Wetzler might be part of your shop equipment.

(Continued)

CLAMPS AND OTHER DEVICES

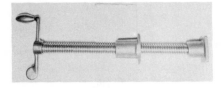

You can use press screws such as this to make your own press, or to make portable or stationary gluing jigs. The pressure plate at the base of screw is removable for easy installation.

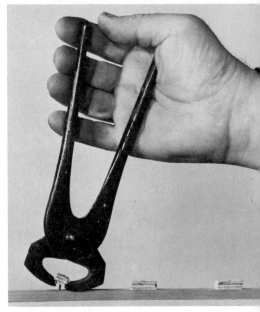

If you lack enough clamps for glued work, parts can be bradded or finish-nailed through small plywood blocks. After the glue hardens, a pincer type nail puller can be used to lift blocks and pull nails. Nail holes can then be concealed with filler.

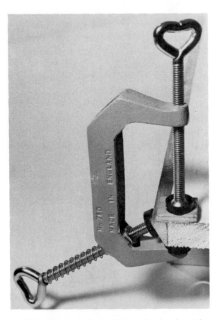

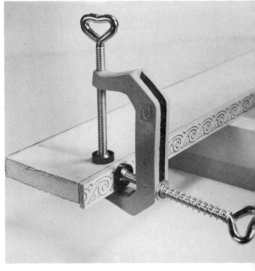

This edging clamp from Anglo-American Distribution Ltd. has two clamping screws for edging work. Angular corners of the clamp permit setting one screw on the surface of the work, and adjusting the second one to clamp beveled edging, as shown left. Slots in clamp permit wide range of clamping angles. At right is the clamping setup for square edging.

Always use scrap-wood pads under pressure points of metal clamps to protect the surface of the work. Judge evenness of clamp pressure by glue squeeze-out. Here the squeeze-out is adequate along the glue line toward the left, but missing entirely at extreme right, indicating that portion of glue line is "starved." Evenness of application is important.

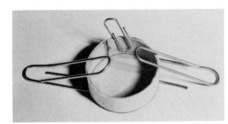

For many small, light jobs, paper clips or spring clothespins can serve as clamps. Here, paper clips hold a cardboard lens cap together while cellulose cement hardens. This lens cap will be used on the camera that took this picture.

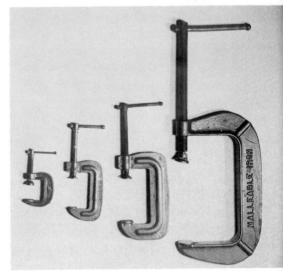

C-clamps are probably the most widely used clamps in gluing. They come in a variety of sizes. Buy the sizes you'll use most.

HOT-MELT GLUING

Hot-melt glue used with this Swingline glue gun bonds a wide variety of materials, including most plastics. Since the glue sets by cooling, repairs are quick. You must keep your hands away from the hot nozzle, however.

Glue-gun adhesive can also be used for caulking. *Caution:* Wear goggles to protect eyes whenever spatter might occur.

Hot-melt adhesive also sticks to masonry. If the work surface is cold, for a stronger bond, warm it with a heat lamp or a propane torch beforehand.

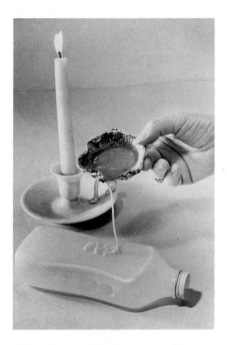

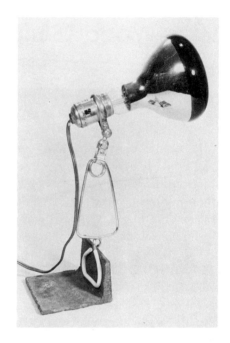

Hot-melt glues are also packaged for use with heat sources other than glue guns. This glue is "Hotgrip" made by Adhesive Products Co. One of the few glues that will stick to polyethylene, it is poured here for the bond of a polyethylene squeeze bottle to a metal bracket. Candle flame was used to melt the glue.

To speed hardening of glues, such as resorcinol, you can use heat lamps. But be sure to check temperature of the wood surface at frequent intervals to avoid fire. If the wood becomes too hot, move the lamp farther away. A short section of heavy angle iron, like this, makes a good stand for a clamp-on lamp bracket.

INJECTOR

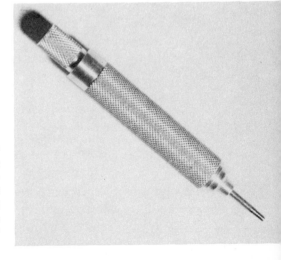

This glue injector has a slender injector tip and a removable piston plunger. You fill it with glue while the plunger is removed. Then insert the tip in the loose joint to be glued (or into a drilled hole in the joint) and press down on the plunger. The gun is available from Albert Constantine and Son.

4 | Wood Gluing Methods

PROPERLY USED, most modern wood glues are stronger than the wood you're likely to use them on. But your gluing methods and the temperature of the work area are major factors in attaining that strength. Since there's often a variation between brands of a given glue, and even between different versions of the same basic type made by the same manufacturer, your first step should be to read the manufacturer's instructions.

TERMINOLOGY OF GLUE TIMING. Gluing instructions for major jobs usually include a timetable covering several different working temperatures. The first item on the timetable usually is the pot life, or working life, of the glue, which is the time (usually in hours) the glue remains usable after it has been prepared. The second item on the timetable is the maximum open assembly time. This, usually in minutes, is the period after the glue has been applied, before the parts must be brought together. Next, is the *maximum* "closed" assembly time—the period after the parts are brought together, but before clamping pressure is applied. The final item is the *minimum* curing period (usually in hours) required before the glued object can be handled. Some instructions specify the pressure period required, as well as the type of handling and finishing work that may be done immediately afterward. Although these timetables vary with the brand, the sample timetable that follows is typical for plastic resin glues.

Temperature (°F)	70	80	90
Pot life (hours)	4–5	2½–3½	1–2
Maximum assembly times			
Open (minutes)	15	10	5
Closed (minutes)	25	15	8
Pressure period (hours)	12	8	5

Note: Assembly time for ready-to-use aliphatic and PVA (white) is about 5 minutes open, 15 minutes closed.

Following the pressure period or minimum curing time, most wood glues continue to gain strength during a "maturing period," which can be a considerable length of time. A typical resorcinol, for example,

42

reaches *ultimate* strength and waterproofness in about 6 days at room temperature. But you can usually begin machining, shaping, and finishing on the project after an 8- to 24-hour period. To gauge the glue's degree of hardness, make a sample joint of scrap wood at the same time as you glue the project, and allow a little excess squeeze-out so you can test hardness. Of course, at higher room temperatures, the tempering period is reduced.

In general, there's no need to rush the assembly if you remain within the maximums specified by the manufacturer. Instructions for glues such as casein, plastic resin, and resorcinol, in fact, may specify a minimum assembly time as well as maximum. The minimum assures that the glue will have time to penetrate the wood pores (to assure bond strength) and to thicken slightly without surface-drying before application of pressure. If no minimum assembly time is given, and you're working with very dense wood or nonporous material, you can keep the joints open during a major part of the *maximum open* assembly time. With resorcinol, this might be 3 minutes. But watch it, because the drying rate of a glue depends on variables such as air humidity, absorbency of the wood, and the wood's moisture content. Don't allow the glue to become overly tacky or to "skin over" before you bring the parts together. If the instructions do not include the information you need, contact the manufacturer for further details. Manufacturers often have special literature for guidance on jobs in which peak glue performance is critical. Yet, for most shop work, you'll find the instructions on the containers good enough.

SOLVING AIR TEMPERATURE PROBLEMS. Temperature in the work area is more likely to be a problem in winter than in summer. If you can keep your workroom at 70°F or more, you can use any type of glue. If you can't maintain 70°F temperatures, you can either select a glue that will harden at the expected temperature, or use some form of auxiliary heat during the gluing and hardening periods. Low-temperature cabinetwork glues include casein (such as National Casein Company's No. 30), which hardens at any temperature above freezing, and aliphatic resin glue (such as Franklin Glue Company's Titebond), which can be used at temperatures from 45°F up to 110°. White glues usually require a minimum 50° temperature. Nitrocellulose cements, such as Ambroid, set by evaporation rather than chemical action and so can be used at temperatures below freezing (though sub-zero cementing isn't recommended). Ambroid cements are limited to smaller gluing areas than casein and the aliphatics because of their quick-drying qualities. At low temperatures, the clamping and curing times of all glues are much greater. For example, at 40°F a typical casein glue would require about 20 hours clamping time and another 96 hours curing time to develop handling strength. At 80°F, clamping time for the same glue could be shortened to about 2 hours, with about 24 hours curing time. If you're using casein, do your mixing in a warm area (not with warm water) in order to assure that the glue is completely dissolved before use. (Do not mix casein in an aluminum container.) With nitrocellulose cements, try to apply the cement in a warm place. The work may then be cured in a colder area. If the cement is applied in a very cold area, increased viscosity caused by the low temperature may make it difficult to spread.

AUXILIARY HEAT. Since low-temperature gluing introduces time-related variables such as increased time for evaporation and absorption, you may prefer to avoid complications by using auxiliary heat.

In northern climates, many tool-rental companies offer auxiliary oil-fired heaters (electrically driven) with outputs as high as 155,000 Btu per hour—enough to heat a house under construction or an unheated garage. If you need only a moderate heat boost, as in some basements or attached garages, you can use either a portable kerosene heater, providing up to 11,000 Btu, or a 110-volt, 1500-watt electric heater that provides about 5,100 Btu. If not stocked by local hardware outlets, the portable kerosene type is available from several sources, including Perfection Products Co., Heater Division, Waynesboro, GA 30830, or from L.L. Bean, Inc., Freeport, ME 04033. Since all of these portable oil and kerosene heaters are flueless, ventilation is extremely important. Whether you buy or rent one, get complete instructions. In any case, first check with local-code authorities for okays.

WOOD TEMPERATURE. When using auxiliary heat for your work area, be sure the wood you'll be gluing is at the same temperature as the heated room. If it's stored in an unheated area, bring it into the heated area at least a day in advance. Be sure to separate the pieces so the heated air can reach all surfaces. Cold wood, even in a heated room, can chill the glue below its required hardening temperature and in many cases, keep it chilled long enough to adversely affect joint strength. To play safe, be sure both the room and the materials are at the required temperature.

In hot weather you can use natural temperature variations to your advantage. For example, I built the all-glued canoes, shown in Chapter 1, outdoors in stages that exploited natural outdoor temperatures. I readied parts for assembly a day in advance and kept them indoors overnight. To extend the working life of the resorcinol glue, I coated and assembled large parts during the cool (70°F) morning hours. Once I brought the parts together and clamped or otherwise held them in contact, I moved the assembly out of shade so the sun would shorten curing time. (At 70°F or more, resorcinol needs at least overnight to harden. Bond strength continues to increase for several days. Check the timetable for the glue you are using.)

PLANNING AND DRY RUNS. Large gluing jobs, as in boatbuilding and major cabinetwork, require advance planning to divide the work into stages that can be done within the glue's time limits. You should also avoid gluing an assembly in such a way that a final part can't be put in

ORDER OF ASSEMBLY

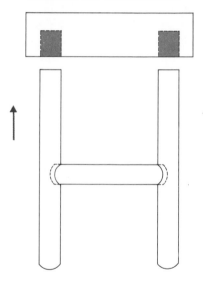

Unless straight table or chair legs are slender and flexible enough to be sprung apart to receive the rungs, they should be inserted into glue-coated mounting holes with rungs glued in place. Otherwise, later the rungs couldn't be inserted.

Angled legs should be inserted with all joining parts glue-coated so that rungs and legs enter mounting holes simultaneously. As the angled legs continue to enter the mounting holes, the rung holes continue to slip over the rung ends.

place without disturbing glued parts. In some cases, too, only one order of assembly is possible. The best way to dodge problems is by making a dry run, without glue. You can even use a dry brush on very long glue lines to make a better estimate of the time needed for gluing and clamping. On boat work, for example, where glue lines may be 16 feet long or more, the dry run lets you know if you'll need a helper in order to coat both joining surfaces before either one dries or skins over. (The major length glue lines of the canoes shown in Chapter 1 were two-man jobs. The method: Workers started at opposite ends, each coating one of the surfaces to be joined.)

GLUE APPLICATION. On small jobs with glue that's difficult to clean from a brush, it's often wise to use an easily cleaned spatula, a disposable wooden paddle, or a pointed stick, depending on the job. In model work with white glue or cellulose cement, wood paddles and pointed sticks can be reused many times without cleaning. Just whittle off the hardened adhesive. For major home-shop gluing jobs, of course, the paint brush is the commonest application tool. Use a cheap disposable brush for hard-to-remove glue, such as epoxy. Select a width that matches as closely as possible the width of the coating required. For example, use a ³/₄-inch width brush for a ³/₄-inch board edge. Apply the glue as you would apply paint, and if the brush is to be cleaned, do the cleaning as soon as possible after applying the glue. If the parts to be joined are well matched so that full surface contact is assured, coat only one of the two meeting surfaces. If the surfaces are long or varied in contour, as in boat-building, coat both surfaces. The same applies to any surfaces where the fit is at all doubtful.

In cabinet and boat work, apply enough glue so a *slight* squeeze-out shows along the glue line where the parts are clamped. You may want to make a few trial joints with scrap wood. Remember, you don't want so much squeeze-out that it runs down the wood surface. (You can wipe it off, but you're wasting glue, and some types may cause staining.) If you want to give the wood an unpainted natural or stained finish, remove squeeze-out very carefully. You can do this with a cloth dampened with the glue solvent (water for water mixed glues) while the glue is still fresh. You can also let the glue set to a rubbery consistency and shave off the squeeze-out in a long, thin strip by sliding a sharp chisel along it. This must be done before the glue is really hard. Once the squeeze-out is hard, removal requires sanding. For natural finishing, a certain amount of fine sanding is usually required, even though you remove the squeeze-out by wiping with a damp cloth or by shaving with a chisel.

The fine sanding removes any minute squeeze-out leftovers, smooths raised grain, and takes off possible remaining glue film that might cause uneven tones during staining and finishing.

FILLING GAPS. If minor gaps appear in poorly fitted joints, they can be filled with wood filler after the glue hardens. If the gaps are likely to weaken the bond significantly, fill them with a glue-coated wood sliver, while the glue is still wet. If you don't discover the gap until the glue has hardened, you can still use the sliver, bonded in place with epoxy. Most epoxies are good gap fillers and will bond to almost anything, including the hardened glue in a misfitted joint. To speed this type of error-patch, you can also use an acrylic adhesive, such as P.A.C., which can harden in as little as 5 minutes if mixed for fast setting.

WATERPROOF GLUED SEAMS. If you want waterproof seams for wood projects, resorcinol glue is your best bet. (Epoxy can also do the job, but resorcinol is ranked first by many boat builders.)

Apply the resorcinol by brush to both surfaces. If a "dry" area appears along the coated glue line (where overly porous grain absorbed glue more rapidly), apply a second coat to that area. Bring the glue-coated surfaces together while the glue is still wet. Where clamping is impractical, you can use nails to hold the seam closed while the glue hardens. If the nails are to be removed later, first drive them through small plywood squares, as shown on page 38. Examine the seams carefully to spot possible gaps while the glue is still wet. Any gap that can't be closed with an extra nail or two should be filled with a glue-coated sliver of wood that fits snugly. But it's better to avoid gaps in the first place.

For small boats involving glued hull seams, the nails used in place of clamps also serve another important purpose. Even though fairly widely spaced, they make the hull assembly rigid enough so you can roll it over *carefully* to examine seams on the other side while the glue is still fresh enough to allow gap-closing measures. In practice, if the glue coating and clamping has been adequate, you're not likely to have gaps. For this reason, in boat work you're safer with too much glue than too little. Excess squeeze-out wastes some glue, but you can wipe it off. And squeeze-out usually means that you're less likely to have glue-starved areas along the hull seams.

ALIGNMENT. In gluing any large assembly, whether it's a boat or a piece of furniture, align the work while the glue is still fresh. A level floor is a help. In cabinetwork, for example, if you know the floor is flat

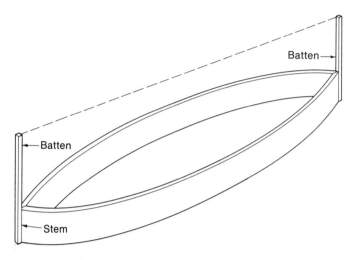

To be sure your boat isn't twisted as you glue it together, fasten an upright batten to each stem. Then stand directly in line with the boat's centerline and align the battens by eye. (Boat shown is a double ender.) If the boat has a transom, fasten one upright batten to the transom mounted vertically at its center, and line up the battens at stem and stern.

and level, you need only assemble a piece of furniture on it to be sure the legs are evenly set. You can check elsewhere with a square. If the floor is uneven, shim up a temporary plywood platform to provide the necessary flat and level surface.

In small boat work, cross beams of 2×4 lumber can be laid flat on the floor and shimmed up level, at intervals along the boat length, perpendicular to the keel. These may be fitted to the hull by blocks nailed on top of them. The object is to prevent sagging, twisting, or hogging (opposite of sagging) of the hull before the glue hardens. The ends of the hull must also be lined up during the gluing to eliminate any twist. This can be done, by eye, after temporarily nailing one batten about 4-feet long (extending upward) to the stem and another to the transom perpendicular to its lateral edge. If the boat's a double ender nail a batten to each stem. If you keep both battens in line (usually possible by eye) as you clamp or nail during the gluing, the hull will line up without twist. On larger hulls, you need heavier uprights, braced perpendicular to the leveled cross support beams, in order to hold the assembly in alignment.

Since resorcinol glue is very hard when set, it may dull edged tools that you use to remove hardened squeeze-out. Power sanding is therefore a better bet for glue removal. A flexible disk sander is good for roughing off the major portion. A belt (fine grit) or reciprocal sander is good for smoothing. And you can perform touch-up by hand sanding. In using the disk sander, be sure the disk is traveling in the direction of the wood grain at the point where the disk contacts the work. This minimizes hard-to-remove cross-grain circular scratches.

5 | Veneering and Laminating

FOR THE PURPOSES of this chapter, the term *veneering* will apply to the process of bonding a thin layer of decorative wood to the surface of a thicker piece, usually less attractive. The object in veneering is not merely to disguise inexpensive wood as a costlier wood. Often, veneering is the only practical means of combining exotic appearance with adequate strength. The reason: Many of the most highly prized grain patterns are found only in the weakest part of the tree. The burl grains of walnut, poplar, and other species actually develop in excrescences of the trunk caused by an injury to the tree. Yet the veneers cut from these burls contain decorative figuring sought for top-quality cabinetwork. Because of the curled and wavy grain, however, wood of this type does not have the strength required for structural parts. So wood for the basic

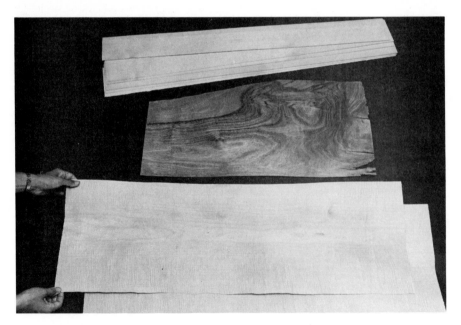

Veneers come in a variety of sizes and shapes, with widths ranging from $1/28$ inch to $1/64$ inch. The curly maple (bottom) measures about 12×36 inches. The pine at top is 4×36. Fancy pieces such as the walnut burl in the center are usually irregular in shape and often fairly small.

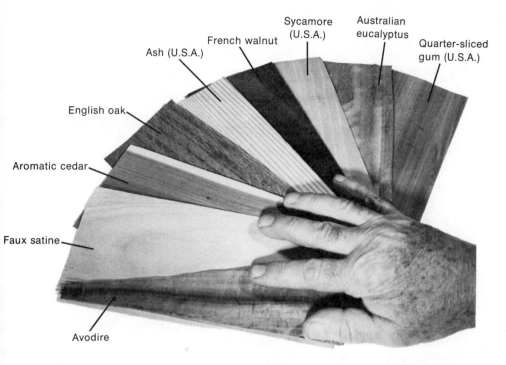

English oak

Ash (U.S.A.)

French walnut

Sycamore (U.S.A.)

Australian eucalyptus

Quarter-sliced gum (U.S.A.)

Aromatic cedar

Faux satine

Avodire

These 4 × 9-inch veneer samples from Constantine give some idea of the tones and grains available. You can buy quantities of samples like these to help you decide which larger sheets you'd like to use. Or you can use the samples themselves for smaller decorative work.

structural part is selected for strength rather than appearance, and veneer is bonded to it for its decorative value.

Weight, too, is sometimes a factor in the decision to use a veneer. Tropical woods such as ebony, with a weight close to 80 pounds per cubic foot, might easily make a large piece of furniture too heavy to handle if used in solid form. So lighter woods are used for the underlying structure.

Methods of veneering. The traditional veneering method calls for clamping the veneer tightly to the glue-coated surface to be veneered. For large areas, a veneer press, like the one shown, is widely used. You can save money by making your own veneer press, using ready-made press screws available from suppliers such as Albert Constantine and Son, Inc., 2050 Eastchester Road, Bronx, NY 10461. The general construction is shown in the photo and drawing in Chapter 3. Overall size and number of press screws may be varied to suit the work planned. The press screws shown in the homemade veneer press photo are made by the Adjustable Clamp Co., 417 North Ashland Ave., Chicago, IL

60622. Small veneering jobs, however, can be done with ordinary C-clamps or handscrews and scraps of plywood, as shown in Chapter 3.

Veneering glues. Though many different types of glue can be used for traditional veneering, one of the most widely favored is urea-resin glue (often called plastic-resin glue). Its advantages lie in its light color, water resistance, high strength, stain resistance, and rapid cure at room temperature. In addition, it is unaffected by the usual solvents in finishing materials. But since plastic-resin glue is not a gap filler type, it requires ample clamping pressure to assure a tight contact over the gluing area, and it must not be used at temperatures below 70°F. Also, it should not be used for gluing oily woods such as pitch pine, teak, and yew, or for gluing wood to metal. Nor should it be used where the bond will be subjected to high temperatures or where you can expect intermittent periods of heat and dampness, as on kitchen countertops or on radiator tops.

Casein glue is a good substitute for urea-resin glue under the unfavorable conditions just described. Casein glue can be used at any temperature above freezing, though setting time is lengthened at low tempera-

In the left-hand photo, tape holds together diamond-matched veneer along juncture lines. The pattern is formed by joining identical veneers cut from the same "flitch." This is a squared timber cut from the original log. You can buy veneer already matched, as shown, or make your own, using identically grained veneers. In the right-hand photo, the tape has been removed to show the matched-grain pattern. Many other patterns are possible, as shown in the drawings that follow.

The herringbone veneer pattern is formed by cutting straight-grained veneer into strips angled to the grain, and then flipping them in alternate strips.

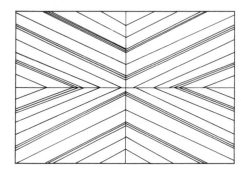

This matched diamond pattern is angle-cut from straight-grain veneer. As a guide in veneer cutting and matching, draw lines on paper, approximating grain type. Then cut the paper with scissors and make experimental matching patterns. Finally, cut the veneer with a fine-toothed saw and joint the edges.

By cutting both across and with the grain of the veneer you can produce four-piece matching like this. Make the kerf (cutting line) as narrow as possible by using a thin-bladed saw.

tures. Because casein glue is alkaline, it tends to saponify the film on the surface of oily woods, permitting the adhesive to bond to the wood fibers. It is also a gap-filler type glue suitable for use where bonding surface irregularities can't be entirely eliminated. Because of its alkaline nature, however, casein glue may discolor woods containing considerable amounts of tannic acid, such as oak, redwood, and mahogany. Such

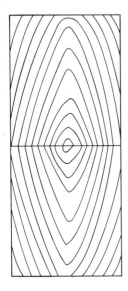

Where grain pattern permits, end-to-end matching gives an interesting appearance like this. But not all grain patterns lend themselves to this treatment. Examine the grain and decide.

Side-to-side matching is easy when you have veneers from the same flitch. Check with your supplier as to the availability of suitable veneers. Also check on the availability of veneers already matched and taped in the pattern you want to use.

stains are likely to occur on thin veneers of highly figured types in which portions of the figuring consist of end grain. The fluid of the glue can penetrate these areas and cause staining. A heavier glue mix (less water), is sometimes spread thinly with a stiff brush to reduce penetration. When staining does occur, it may often be removed with oxalic acid sold in paint stores. Follow the instructions on the container for mixing and using the acid, and protect your hands with rubber gloves and your eyes with goggles. Also be sure that any container in which the acid is kept is clearly labeled *poison*. Don't mix casein glue in aluminum pots.

Gluing methods. Another approach to gluing of oily wood consists of wiping the wood surface with alcohol (such as shellac thinner) before gluing, and then following up with light fine-sanding after the alcohol evaporates. The alcohol wiping should be done in such a way as to lift off any surface film from the wood. Be sure the alcohol you use contains no petroleum denaturant. With the oil film removed, any of a wide variety of adhesives may be used. To avoid problems with a major project, however, try the method you plan to use by applying it to a small test sample made up of the same woods you'll be using.

Here is an important point for veneer gluing by traditional methods with water-mixed glue: *Always apply the glue to the base material, not to the veneer.* Moisture from the glue would otherwise be absorbed by the veneer to a greater degree, causing it to expand considerably. Also, always apply pressure to the veneer as quickly as possible after it has been set in place on the glue-coated base material. Some expansion of the veneer will occur, followed by shrinkage as the glue dries. If you do the job as described above, however, you'll minimize the chance of splitting due to swelling and shrinking of the veneer.

The usual woodworking steps in gluing veneer begin with the preparation of a smooth, clean base surface. After applying glue to this surface, press the veneer (previously cut to size) onto the glue-coated surface by hand, carefully flattening and aligning it. Be sure there is ample glue at the edges. Once you have pressed the veneer into place, fasten it at each corner with a veneer pin. These pins, like very thin brads with almost no head, are available from cabinetmaking suppliers. Standard brads may be used instead, such as ¾-inch No. 20 size or smaller with heads clipped off (length depending on panel thickness). The object of the veneer pins or brads is to keep the veneer from sliding out of position when pressure is applied. After the veneering glue sets, the pins may be driven below the surface with a fine nail set, or pulled out after twisting carefully to free them from the glue. In either case the holes are concealed with suitable filler.

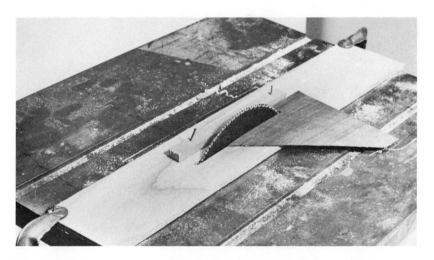

If you're cutting veneer on a table saw, clamp a plywood panel over the saw-blade slot, as shown, to reduce the size of the slot. This prevents shattering of the veneer during cutting. Wood battens bradded to the plywood may be used as guides.

This homemade veneer press employs clamp screws made by Adjustable Clamp Co. and available from cabinetmaking suppliers.

Once the veneer is pinned or bradded in place, with the upper ends of the pins or brads about ¼-inch above the veneer surface, place a double layer of newspaper on top of the veneer, and place a piece of plywood, called a *caul*, on top of the paper. The caul should be a little larger than the veneer surface. Push the caul down by hand firmly enough to force the projecting ends of the pins or brads into it. Then, if both sides of the panel you're working on are to be veneered, turn the panel over and repeat the whole procedure on the other side. You can then put the entire assembly into your veneer press or other clamping system, and then tighten up. The sooner you apply the pressure, the better. If you're using casein glue on veneer that's likely to show stains, remove the work from clamping pressure after no more than six hours, assuming the working temperature is high enough to stiffen the glue by that time. If there's no risk of staining, leave it under pressure overnight or longer.

If the veneer is in sections held together by paper tape, as it is in inlay designs and grain-matched forms such as diamond patterns, it is always applied to the glue-coated base surface with the paper tape on the exposed surface — not the glued surface. The tape is removed after the glue has set.

When a large panel such as a tabletop is to be veneered, you can use plywood or particleboard for the core. But if you are making up your own core from regular lumber, the lumber should be in relatively narrow strips, edge-glued to make up the required area. The strips should be arranged so the inner and outer faces of the strips are alternated, as shown in the accompanying drawing, to minimize distortion. The drawing of the cross section shows the reason for alternating strips. The cross bands are most often made from Honduras mahogany or poplar because these woods resist curling, and are easily glued. Crossbanding veneers of this type are available from veneer dealers and cabinetmaking suppliers. The overall construction shown is termed *balanced construction* because the same number of veneers with matching grain directions (crossband and face veneer) are used on each side of the core. This results in a highly distortion-resistant panel, but requires a considerable amount of work.

Veneering without clamps or press. Often there's no need to follow traditional veneering methods. For example, there's no need for total au-

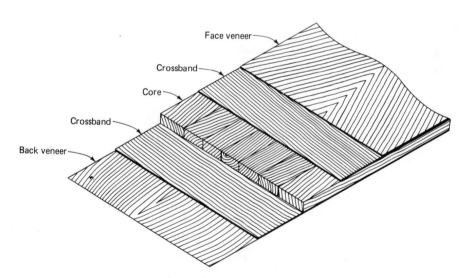

If you make the complete assembly for a large, veneered panel such as a tabletop, layering should be like this. Note the narrow core strips edge-glued with grain of alternate strips reversed. Crossbands at right angles to the core bond it tightly and minimize distortion. Back veneer runs parallel to the face veneer.

thenticity in all antique restoration. So it's possible to do your veneering with the same basic type of contact adhesive used to bond plastic laminates to countertops. For bonding of wood veneers, however, the consistency of the cement is modified in manufacture to promote even coating. Otherwise, unevenness that would not show through plastic laminate might show through very thin wood veneer. Constantine's Veneer Glue is a contact cement of this type. It is usually applied by brush to *both* meeting surfaces—the base, or core, *and* the veneer. As in plastic laminate work, both surfaces are then allowed to dry completely. The surfaces should have a uniformly glossy appearance when dry. If the wood of either surface absorbs more cement in some areas, leaving them dull, apply an additional coat. Allow ample time for drying, typically around 30 minutes or more. (The cement usually retains its tackiness for 3 or 4 hours after drying, but surfaces should be mated sooner.)

As in plastic-laminate work, place a paper *slip sheet* on the base surface after the cement has dried. This sheet must cover the entire surface. Then lay the veneer on top of the slip sheet and align it precisely with the base. Next, preferably with a helper holding the veneer in alignment, carefully inch the slip sheet out from between the two dry, cement-coated surfaces. Move the paper out an inch or so, making sure the veneer and base are aligned, while pressing the veneer down on the base where the paper is no longer between the two surfaces. Once the two surfaces are in contact with each other they bond instantly, so no adjustment or realignment is possible. Held by the initially bonded area, the remainder of the veneer can be pressed into contact with the base as you completely remove the slip sheet. Use care to assure that the veneer is flat and free of wrinkles as you press it into contact with the base. A small rubber roller (such as a photo print roller) is a good tool for pressing the veneer firmly in place.

Veneer cement is rubber-based, so you won't encounter moisture absorption problems as with water-mixed glues. Hence, "unbalanced" veneering is usually practical, with resultant savings in work and expense. Also, because it adheres to a wide variety of materials, including metal, it can be used to apply veneer over old surfaces, including those already veneered. Just be sure the surface is smooth, as well as free of dust, wax, and furniture polish. Also be sure that holes or digs are filled and sanded level. It is even possible to veneer a metal filing cabinet, or a tin can. You can use ordinary contact cement for smaller projects. But for larger veneering projects, it's better to use a contact adhesive made especially for wood veneer. As an example of a small project, note the inlaid jewel chest, shown in photos on accompanying pages, made from

(*Text continued on page 63.*)

VENEERING A SMALL BOX

1. What you can do with veneer and contact cement is illustrated using this dried codfish box. The box is shown just as it appeared after the codfish were removed.

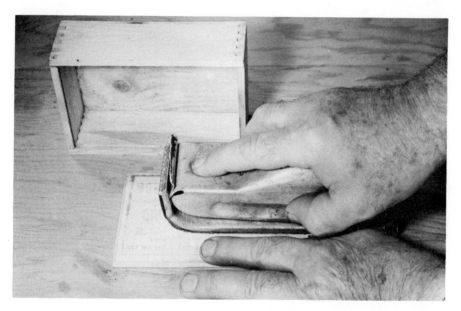

2. First step is to sand off lettering (which here is recessed) using medium-grit aluminum-oxide paper on a hand sander. The sanding paper also works when wrapped around a wooden block.

3. These are just a few of the ready-made strip materials you can buy for use in veneering and surface-decorating jobs. Some are simply flat inlays. Others are carved or indented strips. These shown are from Constantine. *(Continued)*

4. Here an inlay strip has been glued to the end of the box. The strip for the long side of the box rests on top. A multiple-blade knife (right) is a good tool for close trimming. Side-cutter pliers simplify roughing to size.

5. Inlay designs like this one are usually made with veneer around them for protection. You must cut away the veneer to free the design for use. A trimming knife, as shown, is a good tool.

6. With much of the protective surrounding veneer cut away, the inlay design looks like this. Note: It's easiest to cut away the protective veneer in sections, as shown. The inlay is shown with taped side down.

7. With excess veneer cut away, the outline of the inlay can be traced onto the veneer sheet like this. (There are other ways too.) Use a sharp pencil and keep the line close to the inlay. The inlay is shown taped surface facing up. This will become the exposed finished surface after the tape is removed.

8. After you cut out the traced outline of the inlay with a trimming knife, coat the underside of the surfacing veneer with contact cement. The box lid under the bottle of cement will also be coated. After the veneer and box top are dry, press them in place on the box top. Cement-coat the inlay and place it into the cutout.

9. The veneer and inlay are here being rolled with a piece of pipe, to assure firm adhesion and flatness. Next the lid surface should be given final smoothing with 400-grit aluminum-oxide paper.

10. Once veneered, the old codfish box becomes an exotic jewel chest. The inlaid lid slides open as it did originally.

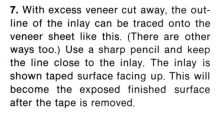

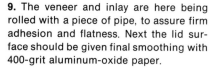

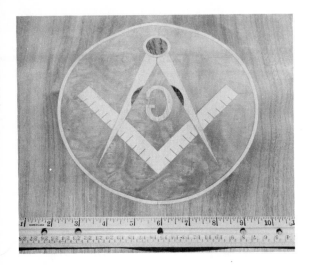

Emblems of many fraternal organizations are available in inlay form. This is a Masonic emblem, shown from the back because its front surface is still covered with tape.

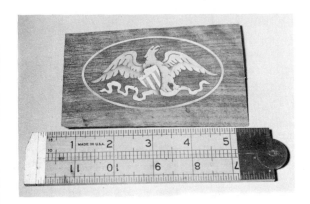

This is a small American eagle inlay, also shown in reverse from the back.

These are samples of plastic Micarta laminates. For small projects, they are contact-cemented by the same method shown for the jewelry chest on preceding pages.

Glue-on molded wood carvings are another form of ready-made cabinet decoration available from cabinetmaking suppliers. Most have predrilled brad holes. Brads hold the carving in position while the glue hardens. White glue or aliphatic glues are often used.

a dried codfish box. The inlay patterns are available from cabinetmaking suppliers, such as Constantine's.

LAMINATING. In this chapter, we'll define laminating as the process of gluing together a number of relatively thin layers of wood to make a heavier piece. For some purposes, as in the construction of wooden airplane propellors, canoe paddles, and large structural beams, the wood layers are assembled so the grain direction of each layer is parallel to that of the others, for maximum strength.

Plywood, on the other hand, is made up so that successive layers have grain at right angles. This results in a dimensionally stable, distortion-resistant panel, with "two-way" strength. (The same basic crisscross grain principle has long been employed by means of "crossbands" in veneering using traditional methods.)

Curved-wood laminates. To produce curved wood forms in the home workshop, laminating is usually a fairly simple procedure. Still, the details and methods vary with the shape you want to produce. Typically, the work is done on a "jig" which is merely a temporary form, often made from scrap wood, to hold the laminated piece in shape while the glue hardens. But you have to plan so that you can set the glue-coated laminations in place, usually one at a time. It's also essential that the jig be covered with at least two layers of paper in such a way as to prevent the work from being glued to it. If you use a single layer of paper the glue may penetrate it enough to form a bond between the two wood surfaces.

This is a scrap-wood jig used for laminating boat coaming with resorcinol-resin glue. In this jig, blocks have been clamped to the plywood backboard to serve as pressure points to produce the desired curve. A cardboard template is used to gauge the curve before clamps are applied to the laminated part. Clamped pressure-point blocks permit adjustment of the curve to match the template. Newspaper prevents the work from sticking to the backboard.

This is the laminating jig for the outrigger beam of a trimaran sailboat, shown on page 65. The large C-clamps are used as posts around which laminations are bent. The smaller clamps are used to pull the laminations together. The glue is resorcinol. And the work is done on a 2 × 8 as a base.

A simple curve, such as a shallow arch, can be formed on a jig with as few as three pressure points, as in the accompanying photo, if the wood is allowed to take a natural bend. It's wise, however, to have a number of C-clamps ready because of possible variations in the bending of individual strips. Complex curves, such as the gull-wing form of the canoe outrigger beam, require a greater number of pressure points. Since reverse curves are involved, the C-clamps may be required in larger number.

In any home-shop laminating, a dry run is an important preliminary after the jig is completed. Slip the laminations in place without glue, noting how you will apply the glue when you do the final job. To allow for possible unevenness of fit, it's best to coat both meeting surfaces, and to avoid any assembly procedure that requires hand bending of a number of laminations at one time, even while the glue is fresh. This can be difficult. For most home-shop laminating involving work of moderate cross-sectional size, large common nails (such as 6-inch) driven into a jig of nominal 2-inch lumber serve well as pressure points. The nail heads should be hacksawed off, and a drilled wood block should be used to assure that they're driven perpendicular to the wood. The headless nails simplify removal of the finished work from the jig, and permit its reuse if more than one laminated piece of the same type is to be made.

The smaller the radius of the curves you want to make, the thinner the individual laminations must be. Inch-wide strips of $1/28$-inch thick veneer, for example, have been rolled into 2-inch diameter, 7-ply, glue-

The completed outrigger beam, shown in its jig on the bottom of page 64, has a broad gull-wing form and will retain the form permanently.

In the making, this paddle consists basically of one ³/₄-inch white-pine board sawed to paddle shape with a saber saw. The shaft is shown being laminated with ¹/₂-inch strips on each side. Plastic resin is the least conspicuous glue, but resorcinol provides peak waterproofness. Later, the paddle will be shaped further by means of tools such as draw knife, block plane, rasp, and disk sander.

bonded mast hoops for small sailboats. The thinnest lumber generally available at lumberyards is pine lattice strip about ¹/₄ inch thick, with a usual bending radius of about 30 inches. And, of course, you can laminate layers of plywood. The dry-bending radius usually attainable for the common thicknesses are as follows:

Plywood thickness in inches	Bending radius across grain, in feet	Bending radius parallel to grain, in feet
¹/₄	2	5
⁵/₁₆	2	6
³/₈	3	8
¹/₂	6	12
⁵/₈	8	16
³/₄	12	20

Somewhat shorter bending radii can be made by wetting the plywood, but the plywood must be redried before gluing. For tighter bends, cut very thin strips of regular wood on a table saw and sand them smooth before gluing. In any laminating work, first make some dry trial bends

Curved laminated rafters supporting this church roof evidence the heights to which laminating craft can be taken.

of the material you'll use. Also make a glued-up test section of the same radius and dimensions with the required cross section.

Other laminating. Not all laminating involves curved forms. Heavy, straight structural beams and girders, for example, are often built up of layers of lumber of smaller cross section. One advantage here is that the smaller lumber can be properly dried more readily before laminating than can large solid timbers of the same cross section. The smaller lumber is also less prone to seasoning flaws such as checks and shakes (cracks or separations in the wood). Also, by staggering the joints in laminations, readily available lengths of lumber can be combined for greater overall length. (The same method can be used with curved laminated forms.)

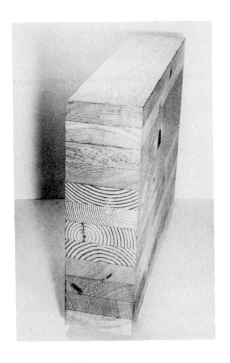

This is a section of laminated structural beam. Note the arrangement of the grain in adjacent pieces. The result is a minimum of distortion.

Laminating glues. If your laminated work is curved or likely to be subjected to sustained loads, do not use glues that may "creep" under either condition (such as most PVA white glue). If a waterproof bond isn't required, casein glue is a good choice, proven in use. It has the advantage of being usable at any temperature above freezing. Where a waterproof bond is necessary, the usual choice is resorcinol, which should not be used at temperatures below 70°F. Either type may be brush-applied. See Chapter 2 for additional specifications.

6 | How to Use Stick-to-Anything Glues

THE VARIOUS "stick-to-anything" adhesives actually stick to *almost anything.* Some set as flexible as a kid glove. Others set hard with a strength measured in tons per square inch, and they do it so fast that you can hold the parts together while the adhesive hardens. Others are slow setting, allowing you plenty of working time on large jobs and long glue lines. To pick the best stick-to-anything adhesive for the job you want to do, read the rest of this chapter. Then, before you buy your adhesive (if it's available from a retail outlet) read the label and instructions. The reason: Adhesive formulas, even for a specific brand and type, may be changed occasionally, sometimes with resultant changes in some of the adhesives' characteristics. Finally, for critical work, try the adhesive on a scrap of the type of material to be bonded.

Epoxies. These are among the best-known stick-to-anything types but must be selected according to the requirements of the job. In general, those with longer setting times are likely to have higher water resistance. Completely waterproof types may call for overnight setting at room temperature. Where it's difficult to keep parts in firm contact, as where clamping or bolting isn't feasible, consider using small dabs of quick-setting epoxy to "tack" the parts together while the longer-setting adhesive hardens. You can also "tack" parts together with dabs of hot-melt glue from a glue gun. This sets in 60 seconds. (See Chapter 7 for details.) If the parts to be glued fit tightly, the epoxy need be applied to only one of the meeting surfaces. Bringing the parts together will transfer it to the other one. Where the fit is not tight, apply the adhesive to both surfaces. If you're gluing two nonporous materials together, as with metal to metal, favor a 100-percent-solids epoxy. This solidifies without need for the escape of solvents, which are sometimes used to produce a desired consistency. As in all gluing, the meeting surfaces should be clean and dry. Epoxies, in general, tolerate faint traces of oil better than most adhesives. But don't assume this lets you be careless about cleaning the work.

In decorative work, as in repairing ornamental china, pick an epoxy that doesn't conflict with the color of the work. Twin-tube epoxies are often colored so the blending of the two colors serves as a guide to thorough mixing of the components. If you plan to use this type, let a mixed sample set before you tackle the job, so you can see the final adhesive color and compare it to the color of the object to be repaired. If it's a bad match, use a clear epoxy. This is available in twin tubes, and piston dispensers (such as Devcon's "Dev-Tube") for small jobs, and in cans for large ones. Be sure the type you buy is labeled "clear." In all epoxy work remove squeeze-out immediately. Have the tools ready. Epoxy is a good choice for strong, rigid (though not brittle) bonds between like or unlike materials.

Water-phase epoxy. These, such as Dur-A-Poxy No. 200 (see Chapter 2) developed as a coating, can also be used on a wide range of materials as an adhesive with unusual qualities. In many instances, it can even be used to bond porous materials to damp surfaces. A fiberglass strip I bonded over a plywood deck seam (merely mopped dry after a heavy rain) was still firmly bonded six years later despite winter snow and ice, and direct summer sun. The epoxy was applied by brush in a stripe slightly wider than the fiberglass strip, with the plywood seam centered in the brushed-on stripe. The fiberglass was then laid in the wet epoxy and pressed down firmly by hand (with rubber gloves). Dur-A-Poxy was then brushed onto the fiberglass for a soak-through coat. The following day a final smoothing coat was applied. A few days later, the entire deck was painted with latex deck paint. The seam has never leaked. Dur-A-Poxy can also be mixed with portland cement, following the manufacturer's instructions, and used then as a bonding crack filler in concrete repair. A final plus: Tools, brushes, and containers can be washed clean with water before the material sets.

Acrylics. These, such as P.A.C. and 3-Ton Adhesive, are not as well known as the epoxies. But they can do the same type of stick-to-anything jobs. They're the type to use for a high-strength bond (tons per square inch) between porous or nonporous materials when quick setting is required. A stiff mix can set in as quickly as 3 minutes. A wet mix can extend the setting time to around 20 minutes. Check the manufacturer's mix-time instructions. Both P.A.C. and 3-Ton Adhesive are excellent gap fillers. Since they are based on a formula originally developed for filling teeth, they set hard, rigidly, and completely waterproof regardless of setting time. They're also unaffected by oil or gasoline. To mix small-job quantities, place the powdered component in a screw-on soda bottle cap with gasket removed, and add the liquid component with a glass dropper, mixing with a thin wood strip until you have the consistency you

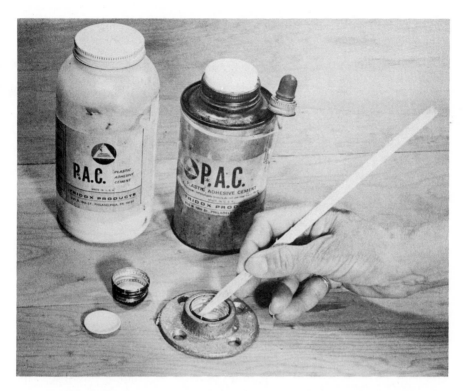

A flange for nominal 1-inch threaded pipe serves as a weighted base for screw-on soda-bottle caps used to mix glue for small jobs. The glue here is two-part P. A. C. acrylic. Expensive, it's used for jobs that need high strength and quick setting.

want. Then apply it to the work with the same stick, and press the parts together. For larger jobs use the same basic mixing procedure with a larger mixing cup and a deep measuring spoon instead of a dropper. If the job can't be clamped, use a fast-setting stiff mix so you can hold the parts (3 to 5 minutes) until the adhesive sets. If any gap shows after hardening, fill it with the same adhesive because it bonds to itself.

Acrylonitrile. These adhesives, such as Pliobond, offer stick-to-anything qualities plus the flexibility of a kid glove when set. Hence, this is the type to use for bonding flexible materials such as fabric, as encountered in quick repairs on sails, convertible tops, and even work clothes. Fabric joints made with acrylonitriles have shown greater strength than stitching. A patching strip about 1½ to 2 inches wide on the concealed side of a fabric rip is typical in repair work. For jobs involving such areas, buy the adhesive in a can or jar with an applicator brush, and apply the adhesive liberally to both meeting surfaces. (This applies to any of the 3 bonding methods. Always try the adhesive first on a small inconspicuous area of the fabric to see if the solvent causes any adverse effects. See Chapter 2 for the most suitable of the bonding methods.)

The adhesive can also be used to bond rigid, porous or nonporous materials, by appropriate method, and combinations of materials. (A boat icebox gasket I bonded to the stainless steel skin of the icebox with Pliobond was still firmly attached 10 years later.) Though not recommended as a wood-to-wood glue, acrylonitriles bond well to wood and can bond most other materials to it.

Latex combination adhesives. These are made in different formulas, some as flexible as the acrylonitriles mentioned above. Patch-Stix (see Chapter 2) is one well suited to flexible fabric bonding and jobs such as upholstery repair. When set, the adhesive is actually a form of synthetic rubber. (A 6-inch rip I repaired with Patch-Stix in the front-seat upholstery of my car was still intact three years later when I sold the car.) The method used depends on the job. For the car seat, I spread the rip open so I could apply adhesive to the inner surfaces adjacent to the rip on both sides. I then inserted an adhesive-coated strip of heavy muslin, and pressed firmly on the adhesive-coated upholstery adjacent to the rip in contact with it, working the rip edges as close together as possible. I inserted straight pins to hold it. Setting time was relatively short.

When the insert strip method isn't feasible, the upholstery can usually be partially removed to provide access to the inner surface of the rip. The strip can then be applied, using the adhesive to bond it in the same way. Once the bond has set, the upholstery can be fastened back in place.

Rubber base. These adhesives (styrene butadiene), such as Black Magic Tough Glue, are stick-all types with high gap-filling qualities that permit bonding rough materials such as brick, concrete, and loose bathroom tiles to porous surfaces. The same adhesive also bonds metal, wood, glass, fabric, rigid plastics, and many other materials. When set, it is rigid but not brittle. To use it, apply it to *one* of the meeting surfaces, "working" the surfaces slightly as you bring them together in order to break the surface film, which forms quickly when the adhesive is exposed to air. To get a quick initial bond, pull the parts apart several times, then press them together permanently. Clamping or propping is necessary only where needed to keep the parts from slipping out of position. For maximum strength, allow about 24 hours. You can remove the adhesive "squeeze-out" with lighter fluid in the early stages or trim it off later.

General information. Many adhesives not regarded as "stick-to-anything" types nevertheless bond to a great variety of materials. The container label often lists materials an adhesive will bond. This is particularly important when working with plastics. If you don't know the type of plastic you're working with, place a drop of the adhesive on it in an inconspicuous place, and remove it before it stiffens. If it has etched the surface, the chances are it will bond the material. In any application *always* read the instructions before you apply the adhesive. Cyanoacrylates, for example, set almost instantly and have high strength, *but* if you read the instructions of a typical brand you'll find the bond should not be disturbed for 1 or 2 minutes. Maximum strength doesn't develop for 12 to 24 hours. Also, you'll probably find a caution *not* to use the adhesive for bonding rear view mirrors to auto windshields, as well as a caution that bonds made with glass may weaken with age.

7 | Hot-Melt Glues

WITH HOT-MELT GLUE you can put a project to use about 60 seconds after you glue it together. And the glue, which bonds to almost any material, is also a gap filler. In the form most widely used in the home shop, hot-melt is polyethylene based, made in cylindrical cartridges that fit electric glue guns. Operating details vary with the make and model of the gun but typically require a gun warm-up of 4 to 5 minutes to reach the glue's melting temperature of 300° to 500°F. (*Caution:* To avoid being burned keep clear of the hot glue and the gun nozzle.)

For new work and repairs, joints must be assembled and pressed to-

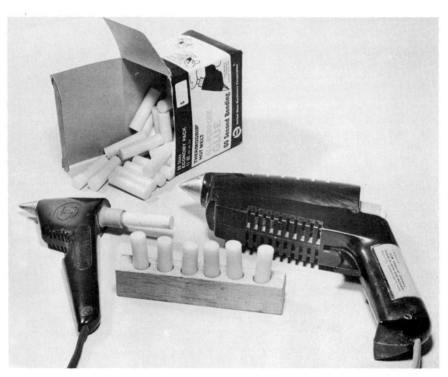

Here are two USM glue guns after 8 years of use. In the foreground, a piece of scrap wood drilled with ⁵/₈-inch diameter holes is a quick-access holder for glue sticks.

gether within 20 seconds after the glue goes on. The glue is applied in a bead from the gun to one surface only, to save time. It should *not* be spread out in a film with a spatula or other tool because this would otherwise speed cooling. When the parts are pressed together, the bead is spread over the meeting surfaces automatically. Hold the parts in position for another 20 seconds, and the glue stiffens enough to hold them permanently. It reaches 90 percent of its full strength one minute after application and develops the remaining 10 percent overnight. Final strength is typically in the range of 200 to 300 pounds per square inch. Although this is of lower strength than that of most of the slow-setting glues, it provides about a quarter-ton holding power in a 2-square-inch area—plenty for the usual cabinet and workshop jobs. As to the seemingly short working time of 20 seconds, look at the sweep second hand of your watch for this period of time and you'll see that the 20 seconds is long enough to perform numerous manipulations. In fact, a 10-second assembly time is enough for most joints. But make a "dry," test assembly to gauge the time you'll need.

You can preheat the surfaces to be joined, using a flatiron or heat lamp, or even a reflector spot or flood bulb. For some small-joints, the hot nozzle of the gun can do it. The extra heat in the joint (especially in woodwork) prevents the glue from chilling as it spreads when the parts are pressed together, and allows more penetration and a better "bite." The gluing surfaces should be hot, not merely warm. In shop-sample wood joints made with and without preheating, the preheated ones show appreciably greater strength, often enough to break the wood when tested to destruction.

To bond a long glue line with hot-melt gun glue, you must do some planning. In woodwork, the edges of the meeting surfaces can be bevel

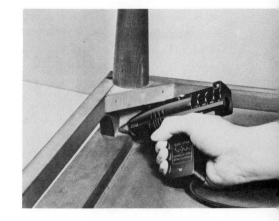

A Swingline gun here applies hot-melt glue to a loose rim and corner block on the underside of a tabletop. When glue can be applied to the outside of a seam, as shown, a long glue line presents no problem because parts are already assembled. Use the hot nozzle of the gun to flow and smooth glue into corners.

This gun injects hot-melt glue behind the lining of an attaché case to bond it back in place. The gun can also be used in upholstery work, as in bonding gimp to fabric or to wood furniture frames.

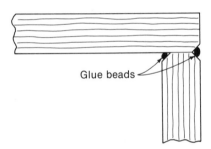

Glue beads

You can assemble a seam with beveled corners, as shown, applying the glue after the joint is assembled. On long seams, it's all right if the first part cools before the final part is gunned. Then you can go back, using the hot nozzle to remelt the glue, smoothing it into the recesses.

planed to form a V-joint, as shown in the accompanying drawing. The bead of hot glue can then be applied in the V after the parts are assembled. By passing a hot flatiron over the glue bead in the V, you can remelt it and flow it against the surfaces. If the glue doesn't completely fill the V, just flow in another bead and repeat the procedure. If the glue deposit ends up high (raised above the surface) don't sand it or else the result will be a rough, stringy surface. Use a very sharp tool such as a razor blade or a keen-edged plane to shave off the excess. The smoothed-off glue can be painted, but not stained by conventional methods. It resists acids and alkalis fairly well. But it has poor resistance to solvents, though brief contact may not cause trouble.

In metal gluing, the bead of glue may be applied to either or both of the meeting surfaces, depending on the glue spreadout required. Then you can apply heat to the outer surface of the metal to remelt the glue already applied to the inner, meeting surfaces. The same method is often used in soldering.

To bond fabric to fabric, flow the glue bead onto one of the fabric pieces. Then lay the other piece on top and iron them together with a flatiron set for full heat. In this and other work involving a flatiron, any glue

This small USM gun is "tacking" a small wood part in place to hold it while slow-drying glue firms up around edges. Tacking is the method to use when high-strength slow-setting glue is bonding a joint that's difficult to clamp.

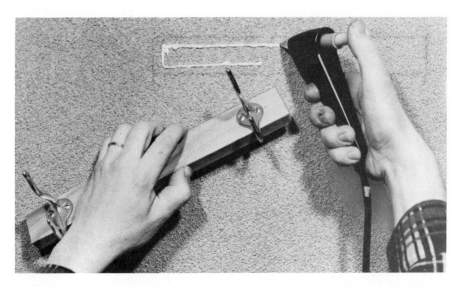

The gun is applying glue to a concrete wall for the mounting of a hat rack. On jobs like this, the rack area should be pencil-outlined on the wall, and then the bead should be applied inside the outline, as shown. For the best bond, preheat the wall mounting area and the mating surface with a heat lamp.

Seams around air conditioners can be sealed with hot-melt caulk. Waterproof caulking can even be used to stop leaks in rainspouts and gutters. The surface must be dry. Preheating helps.

remaining on the surface of the iron can be wiped off with a wad of cloth while the iron is still hot. Always test the fabric for the heat it can tolerate.

In many types of repair work the glue bead can simply be flowed into inside corners, such as loose corners of a drawer, and smoothed to form a fillet, using the nozzle as a tool.

For caulking work, buy caulking "ammunition" for your glue gun. This is available to fit most glue guns. Though formulas differ with the brand, both hot-melt glue and caulking have good bonding strength, but generally should not be subjected to temperatures over 130°F. Because of formula variations, follow the instructions for the brand you use.

8 | Adhesives for Construction and Paneling

SELECTION OF ADHESIVES for home construction and paneling should be based on the job to be done and the gluing conditions. In subfloor work, some adhesives can be applied in freezing temperatures, even on wet or frozen surfaces (though this is not recommended), and can reduce required nailing by as much as 75 to 80 percent. Indoors, as on paneling, the right adhesive can often eliminate the need for nailing altogether. In addition to saving work, those used in subfloor and similar applications increase structural rigidity and eliminate possible future problems such as squeaky floors.

There are important differences between construction adhesives and the conventional glues. First, the typical elastomeric construction adhesive is formulated to remain moderately flexible permanently. If the house settles unevenly or if the framing shrinks after the house is enclosed and dry, the construction adhesive has enough "give" to retain its grip. A glue that sets hard, like casein, even though of much greater strength, may pull off a layer of shrinking wood at the joint and thus lose most of its grip. Hard-setting woodworking glues, however, are commonly very much stronger on a square-inch basis, often 4 or 5 times stronger. So for maximum strength, such as in laminated beams and cabinetwork, they're the choice. But where maximum strength isn't required, construction adhesive is usually the better choice.

CARTRIDGE APPLICATION. When used on house framing, as in bonding plywood subflooring to joists, construction adhesives are usually bought in cartridges and applied with a caulking gun. (Professional builders often use power equipment. This sometimes includes a "wand" or extension tube with a guide that steers it along the joists to assure a centered bead.)

In cartridge application, the tapered cartridge nozzle is cut to a point sized to lay a ¼-inch diameter bead along the centerline of the joist edge. (The adhesive need be applied to only one of the joining surfaces.) Where two plywood panels meet along a joist, two beads of adhesive are applied, one on each side of the centerline. Where panel edges meet at right angles to the joists, they should either be tongue-and-groove edges or should be supported by cross members between the joists. Adhesive should also be used in the tongue-and-groove joints between panels in order to take full advantage of the glue's stiffening effect on the overall structure.

Approximately $1/16$ inch should be allowed between all panel end and edge joints to accommodate any expansion. As to timing, the work should be planned so extruded adhesive is not left exposed for more than half an hour before assembly, unless the manufacturer's instructions indicate otherwise.

When the subfloor is nailed and "walked" into firm contact along the joists, the $1/4$-inch-diameter bead spreads out to a width of an inch or more. Since typical brands like Weldwood subfloor and construction adhesive (on which this data is based) have a strength of about 350 pounds per square inch when fully set, it's easy to appreciate their ability to increase structural rigidity. On a single 12-foot joist (144 linear inches) the bonding strength multiplies to approximately 50,400 pounds, or more than 25 tons. Even if you don't do a perfect job, this gives you leeway, because Weldwood is also a gap filler. In addition, it is nonflammable.

Where a mastic adhesive such as Panel Weld is used to apply interior paneling directly to studding, the bead is run along the centerlines of the studs in the same general manner I described for joists. To speed the escape of solvents, in this case petroleum spirits and ethanol, a temporary pull-apart procedure may be used, as shown in the accompanying photos. To utilize this procedure, drive a small finishing nail part way into the paneling near the upper corner on each side, after the panel is held to the studs by the adhesive. (The nails should also go into the studs.) The nails can then act as temporary hinges, letting you pull the panel out about 6 inches at the bottom, separating the glued surfaces. A pull-apart period of 8 to 10 minutes is recommended for solvent escape. Then press everything back in place with good overall pressure. For best bonding results, the adhesive should be compressed as thinly as possible. Once the panel is back in place you can either drive the top nails in all the way or remove them.

Methods may vary with different brands, so follow the instructions for the brand you're using. Also, heed all cautions. Some types are flammable, some are not.

If the adhesive is used to apply paneling over an existing wall (such as a finished drywall), run the bead all the way around the panel, far enough in from the edges to avoid excess squeeze-out. As shown on page 82, the bead should be about 1 inch from the edges. If you want to be

GLUING PANELING

Most panel and construction adhesives are available in cartridges to fit caulking guns. Shown here is Weldwood Panel and Construction Adhesive.

If paneling is to be applied over studs or furring, cut the cartridge nozzle so that it lays about a ¼-inch diameter bead. Make a continuous line along furring and cross members.

Drive a pair of finishing nails at the top of each panel to keep the panel aligned. Then swing the panel out as shown here for about 8 minutes to let solvent escape.

First, press paneling firmly in place by hand. Then use a rubber mallet and a towel-wrapped block along seams.

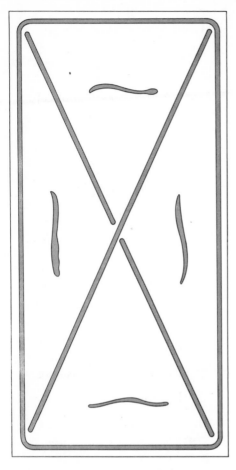

If you are applying paneling to an existing, finished drywall or plaster, apply the adhesive bead to the back of the paneling as shown.

sure, you can try the bead on a scrap of wood and press it against another one as a test. Here differences in the bead diameter can affect the spread of the adhesive. To hold the central area of the panel, apply two diagonal beads of adhesive from corner to corner, crossing at the center. Then add short beads near the centers of the triangles formed by the diagonals. In many cases this is all that's necessary to hold the paneling. In older houses where the walls are not truly flat, however, it's wise to set the panels so their vertical edges are at stud locations so a few finishing nails may be used if necessary to hold the seams even.

Foam insulation. For mounting foam insulation, as in insulating a basement, be absolutely sure the adhesive you use is compatible with the foam. If it's not, it may not only fail to hold the foam, but may also break it down by literally dissolving the bonding area. Usually you can buy such an adhesive from the supplier who handles the foam. One of these adhesives especially suited to bonding polystyrene and polyurethane foams is Big Stick Panel and Construction Adhesive. This, and the other adhesives mentioned earlier, are available from Roberts Consolidated Industries, 600 N. Baldwin Park Blvd., City of Industry, CA 91749. They are sold in cartridges and larger big-job containers for application with notched trowel.

GENERAL GUIDELINES. Since formulas vary, it's important that you read the specifications and instructions that accompany any construction adhesive before you buy or use it. Be sure it's suited to the materials you'll be using it on. Some solvents can have an adverse effect on certain plastics, so if you're planning to mount plastic, try the adhesive on a scrap beforehand. Also, in areas where building inspections are made during construction, check with your local building inspector regarding nail spacing required with structural adhesive.

In all paneling work, it's wise to store the panels for a few days inside the house where they will be used, so they can adjust to the interior atmosphere. Paneling stored in a damp shed at the suppliers, for example, may swell slightly. After a few days of in-house storage, however, it usually comes back to size, because paneling materials usually shrink very little. Still, you don't want it to shrink after mounting, because seams may then show a gap.

In general, nonflammable subfloor and construction adhesives with a water base provide a somewhat longer "open time" before assembly than do types with alcohol or solvent base. This lets you apply the adhesive to more framing parts before applying the plywood. But check the manufacturer's instructions, and don't overextend the time. Paneling adhesives are often formulated for drying times that allow you to mount only one panel per application.

9 | Time and Temperature Tricks

IN GLUE WORK, time and temperature can sometimes play tricks on you if you don't anticipate peculiarities of glues you may be using.

Most popular glues set faster with increases in the temperature of the work area. Hot-melt glues, on the other hand, must be heated before they liquify for application, after which they set by cooling. And some glues set by the evaporation of a solvent.

Yet a few glues differ from those described above by generating a considerable amount of heat on their own. Polyester resin is one of these. Using it, you can avoid problems if you know what to expect as you work. For example, people fiberglassing boats with polyester, have been surprised to find the brush suddenly locked solid into the final portion of catalyzed resin in the can—even though the same resin from the same can is still wet on the boat hull. The explanation is this: The chemical reaction inside the can generates heat faster than is generated in the resin spread thinly on the boat hull, owing to the extensive heat-dissipating area of the hull. You can avoid this kind of trouble if you follow the manufacturer's instructions, and if you use the specified amount of catalyst hardener when you mix the resin. Also, follow the quantity limits and don't mix more resin in a single batch than recommended. Otherwise, heat builds up too fast. Your brush, too, can tell you if your resin is hardening. If it begins to "pull" or stick to the work, lift it off immediately so it doesn't suddenly bond itself in place. If that happens you'll have to cut and sand it off. You can also retard heat buildup if you place the can of catalyzed resin in a bowl of ice. But ordinarily you won't encounter excessive heat buildup except in hot weather.

For special working conditions, such as when you are working in low temperatures or when you need faster or slower curing times, check with your resin dealer or the manufacturer. Large marine supply outlets such as Defender Industries, Inc., 255 Main St., New Rochelle, NY 10801, can often furnish customers with information of this type, specifying in some cases, the amount of hardener to use for fast, medium, or slow curing. (You might want a fast cure for a car-body repair; or you might want a slow cure for a large boat job.) In the event your work must be done out of doors in early spring or late fall, you can also obtain low-temperature polyester for use in the 45° to 55°F range rather than the usual 70° to 75°F range. *Important:* Ask your supplier for any special information you need before you buy resin or begin your job. (*Note:* Polyester is not generally recommended for use on redwood, cedar, oak, or polystyrene foams such as Styrofoam. For these, epoxy does the trick.)

The epoxies commonly used for large jobs, such as fiberglassing, generally react to heat much the same as oil does. Typically, the freshly mixed epoxy shows a lower viscosity (gets thinner and runnier) as it gets hotter. However, if you let it set for a few hours, enough to stiffen it somewhat, the heat usually works the other way, hardening the resin more rapidly. But because epoxy formulas vary widely, always follow the specifics that come with the brand you're using. Typical average curing time at 85°F summer temperature is 8 to 10 hours. At 100°F, 5 to 7 hours.

A glue with a time quirk is casein, although it responds to heat in the conventional manner, setting faster with increased temperature. Its time quirk shows immediately after mixing with water according to instructions. At this point most powdered casein glues have a stiff consistency totally unsuited to spreading. But 10 or 15 minutes later, with no extra water added and no attention, a very brief stir turns the mix to a smooth, creamy consistency that spreads easily and smoothly. The explanation is in the fact that the water first simply sticks the glue grains together, but during the ensuing few minutes, the water actually dissolves them, completely changing the consistency of the mix. *Very important:* Wait for the natural change; do not add water to the stiff mix.

When there's need to have a prescribed workroom temperature range for a glue, be sure you measure the temperature at the level the work will be done. Thermostats and thermometers, for example, are normally mounted at eye level for easy reading. But the temperature at that level may be considerably different from that at the work level. In one instance, I noted a temperature of 65°F on a mounted thermometer inside an outbuilding where I planned to do some gluing. But the temperature at the ceiling was 85°F. At floor level, it was 28°F. Two small electric fans mounted on the ceiling beams to blow downward at the work area brought the temperature into the range necessary for the glue. Another method that can be used with some glued projects consists of assembling the work above floor level, as on the workbench or on a work table made by laying boards or plywood on sawhorses. This raises it out of the chilled zone near the floor. But always measure the temperature at the table height. Some glues, partially hardened below the instruction-specified temperature, do not acquire peak strength even though later raised to the correct temperature.

Some boatbuilders, using resorcinol type glues, invert small hulls on sawhorses and place fire-safe heaters under them so heated air collects inside the hulls to speed the glue hardening. This, of course, must be done with care, and must be watched until the heaters are removed. (Heat buildup can be surprisingly rapid in some instances.) An arrangement of this type should not be left unattended, and should not be used with flammable adhesives. The temperature inside the inverted hull should be measured to be sure it doesn't reach ignition level.

Whether or not auxiliary heat is necessary for the gluing work, it's very important that the manufacturer's labels be followed regarding flammability, skin contact, or toxic fumes. The fumes can be deceptively dangerous because your sense of smell may become less sensitive to them after a time. I once did a fiberglassing job with inadequate ventilation. This resulted in my getting the equivalent of a very bad hangover. And the results could have been much more serious. Yet the fumes hadn't smelled particularly strong. The ventilating methods described in Chapter 10 would have prevented the trouble.

10 | Protecting the Project and Yourself

THE AMOUNT OF CLEANUP WORK to be done after a major gluing job depends to a large extent on the protective measures you take. This pertains to the project itself, to the work area, and your skin.

On the project, you should avoid excess squeeze-out, especially if it can run down visible surfaces of the work. One good method of squeeze-out removal involves wiping with a cloth dampened with the glue's solvent. But the wiping should be prompt and thorough to avoid extra work. Glue that's allowed to soak into the wood is likely to require considerable additional effort if you want to avoid an uneven finish tone, especially if the work is to be stained. Even clear-drying glue can cause an uneven finish because it seals wood pores that would otherwise absorb the finishing stain.

Aside from wiping, there are several other ways to protect your project from squeeze-out trouble. One of the simplest is to use the correct amount of glue for minimum squeeze-out. You can gauge this in advance by testing the glue with scrap wood and clamps. (There should be a slight squeeze-out.) You can also protect the work with cellulose or vinyl tape on the low side of the glue line, or on both sides if the seam is vertical. Apply the tape before gluing, of course, and be sure it's firmly bonded, especially close to the glue line. If it isn't, there's a chance the glue may work under it by capillary action. You can use the tip of a teaspoon to press the tape down for a tight seal. (This method was developed some years ago to prevent polysulfide-type boat-seam caulkings from adhering to the planking adjacent to the seams, because these caulkings are difficult to sand off.) Another method calls for the application of finish coats to seal the exterior of the work before applying glue to the mating surfaces. This prevents soak-in of squeeze-out. Allow the finish to dry before gluing the parts together. Of course, the finish must be one that will not be affected by the glue. In addition, always have a container of glue solvent at hand to clean squeeze-out from the tape.

For most woodworking glues the solvent (before hardening) is water, so it can be in an open top container such as a jelly glass for easy dipping of your wiping cloth. The cloth should be moist but not wet, because you don't want water or other solvent to soak into the glue line. A jelly glass

Tape helps you avoid staining from glue squeeze-out. A teaspoon ensures a firm tape bond to the lower face of wood.

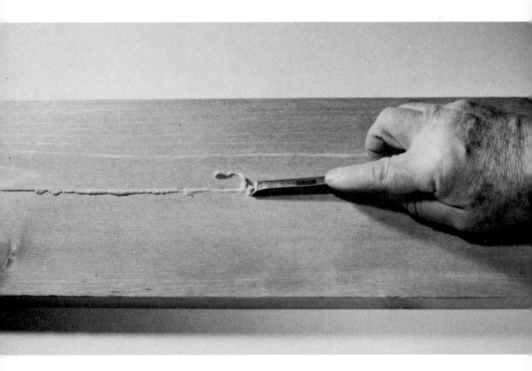

After glue has set, you can often remove squeeze-out with a chisel. First run a test on scrap wood. On finished surfaces, avoid gouging the wood.

is also better than a can for a glue pot, because acidic glues absorb iron from metal containers. The absorbed iron in the glue may then react with woods, causing a nearly black glue line.

To wipe glue at an inside corner, wrap your wiping cloth around your finger or a stick. If the squeeze-out is slight and without runs, you may prefer to wait until the glue sets partially to a soft consistency and then remove it by sliding a chisel blade along it, bevel-side up. When the cutting edge of the chisel rides firmly on the wood, it will curl off the squeeze-out in a continuous shaving.

If you end up with a glue stain despite your precautions, sanding is the final remedy. Often, the corner of the sanding pad on an orbital sander can make the job easier, especially in tight spots. But don't use a coarse-grade abrasive.

BENCH AND FLOOR SURFACES. Protecting the bench surface or floor area is a major concern too. And it's a simple measure. Just lay newspaper over the area, using about three layers, because spilled or dripped glue may soak through a single layer and glue the newspaper to the surface under it. If you have a plastic drop cloth, lay it over the news-papers. This minimizes the chance of soak-through, unless the plastic is in bad shape. If the drop cloth is polyethylene, most glues won't stick to it. Be sure you cover an ample area around the perimeter of your project, and watch for dribbles that miss the covering. A handheld brush, for ex-ample, can often deposit a drip-line of glue beyond the protected area. If you notice the drips soon enough you can wipe them up. If they harden they may become a permanent part of the decor, especially if you are using an epoxy on a cement floor. You can sand glue off, but a blemish usually remains. There are chemical preparations that will remove hardened epoxy, although your local supplier may not stock them. One preparation is MS 111 Epoxy Stripper, made by Miller Stephenson Chemical Company, George Washington Highway, Danbury, CT 06810. You can contact Miller Stephenson for the nearest source, or arrange to have your supplier order it. Follow all instructions and cautions accom-panying the preparation.

BRUSHES. If you're using a glue that's very difficult to remove from brushes, brush cleaning usually isn't worth the time and effort. So your best bet is a supply of cheap, disposable brushes. Be sure to have a disposable container, such as a cardboard box, for the used brushes. If you lay a brush down, you may have to chisel it off. For glue, such as resorcinol, that can be removed easily from tools and brushes before hardening, be sure to have containers of solvent (in this case water) ready to use. Just slosh and twirl the brush against the bottom of the container, then rinse thoroughly under the faucet, and your brush will be ready for the next job.

FLAMMABLES AND TOXICS. When you are using flammable adhesives or flammable cleanup solvents, you need ample ventilation and you can't allow any flame, such as pilot lights or other ignition sources, in the work area. For example, flammable contact or panel adhesive may contain toluol, methyl ethyl ketone, alcohol, and petroleum spirits, and may call for cleanup with petroleum spirits or lacquer thinner — all flammable. While these materials can be used regularly without trouble, caution is necessary. For safety's sake, use the same precautions with all flammable materials that you would with gasoline.

As to required ventilation for toxic and flammable fumes, one rule of the thumb requires an air change at the rate of at least 1 cubic foot per minute for each square foot of workroom floor area, assuming a ceiling of average height. This would amount to 200 cubic feet per minute (cfm) for a 10 × 20-foot workroom. Another rough guide calls for a complete change of air in the room every 5 minutes. Assuming a 7-foot ceiling in your 10 × 20-foot workroom, this would come to 280 cfm. Wall and attic ventilating fans are already available, however, with capacities from as much as 700 cfm to several thousand. Such a fan with enclosed motor can be handy for many types of shop work, both for removing toxic fumes and flammable fumes of painting, finishing, and gluing, or cleanup. For suppliers, check your phone book's yellow pages under "Fans-Electric" and "Fans-Ventilating and Exhaust." Large mail-order houses are another source. Dealers and installers of these fans usually can provide good advice on capacities, as can your local fire marshal.

PHYSICAL PROTECTION. For your own physical protection, always wear goggles when working with any adhesive or solvent in a manner that could spatter it into your eyes. If you have any doubts, use the goggles. As you'll note from the cautions on the containers, many adhesives and their solvents can be extremely injurious to the eyes.

To protect your hands, use disposable plastic gloves from your local paint store. These protect you against possible allergic reactions and also eliminate the task of removing partially set glue from your hands — sometimes a lengthy chore. Use the same gloves for the gluing and the cleanup. Polyethylene gloves are good when lacquer thinner is used, but most disposable gloves will do the trick. If you happen to have old rubber gloves that are on their way to the trash, they'll usually fill the bill if they don't leak. But for cleanup, give them a test dip in the solvent you'll use. Because of different formulas for both the rubber and the solvent there's always a chance of chemical reaction.

As to your clothes and shoes, save an old pair of pants, a ragged shirt, and a pair of old shoes for glue work. You can reuse the whole outfit many times. And it will help you do a better job because you won't be worrying about what happens to your apparel. When you're working with glue or paint, there's not much chance of keeping your clothes and shoes spotless anyway.

11 | Planning Projects

WOOD-TO-WOOD GLUING. Probably the commonest type of home-shop gluing, wood-to-wood gluing calls for consideration of the nature of the wood as well as that of the glue. Joint failures that result from overlooking the characteristics of the wood, for example, are often erroneously blamed on the glue. Yet the pitfalls can usually be avoided, by using the guidelines that follow.

Shrinking and swelling of wood. Caused by changes in moisture content, shrinking and swelling are extremely important gluing factors. To minimize these effects, buy your wood for fine cabinetwork two or three weeks before you begin work. Then store the wood in a part of the house that has the same general humidity as the room where the finished furniture will be. Kiln-dried lumber that has been stored under protective shelter is a wise choice. If the lumber becomes a little drier than the house usually is, as during a hot dry spell, don't worry. In most parts of the country, lumber for furniture is usually kept below 8 percent moisture content prior to the manufacturing stages. Then it is often brought to a somewhat drier condition to allow for a moderate increase in moisture content during manufacture and storage. In the home shop, of course, kiln-drying to precise moisture content is rarely feasible. But you'll usually do well if you use the old timer's trick of storing the lumber where the finished furniture will be used. So long as humidity within the house remains about the same, completed furniture isn't likely to suffer. Drastic changes, as from humid summer weather to dry winter steam heat, produce the cracks and splits.

The cut and the grain direction of the wood. These also have an effect that you should keep in mind. In percentage, the wood you're likely to use may shrink 30 times as much across the grain, from one edge of a board to the other, as it does lengthwise. From green to oven-dry wood, the lengthwise shrinkage typically ranges from 0.1 to 0.3 percent, while shrinkage across the grain may run to 8 percent or more depending on the lumber species and the way it was sawed from the log. What is

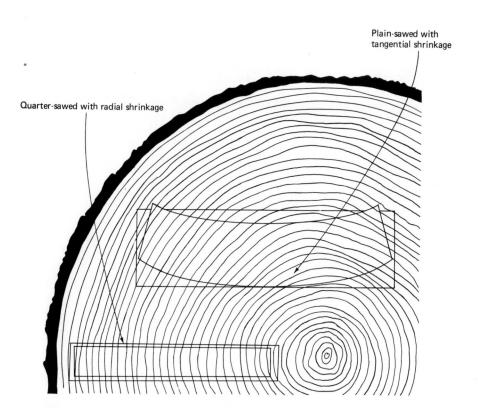

This portion of a log cross section shows characteristic radial and tangential shrinkage and distortion of quarter-sawed and plain-sawed boards. The term *quarter* is used because the logs are first cut into quarters; and *radial* is used because the boards are then cut essentially along radius lines from the log center. The inner face of a *plain-sawed* board touches the log's annual growth rings *tangentially*.

termed *radial* shrinkage is measured on boards cut approximately on a radius from the center of the log straight out to the perimeter — which we usually think of as quarter sawed. *Tangential* shrinkage is measured on boards cut on a tangent to the rings of the log — which we usually think of as plain sawed. Radial shrinkage is invariably less than tangential, and, in general, heavier species (either type of cut) shrink more across the grain than light ones. Plain-sawed white oak, for example, if dried from green condition to 6 or 7 percent moisture content, shrinks about 7 percent. Yet sugar pine, cut and dried in the same way, shrinks only around 4 percent. Radial shrinkage (quarter sawed) is less: about 4 percent for the white oak and a mere 2.2 percent for the sugar pine. (See the table on pages 94–96 for shrinkage of common species.)

SHRINKAGE VALUES OF DOMESTIC WOODS
From USDA Forest Service Wood Handbook: *Wood as an Engineering Material*

| SPECIES | (percent shrinkage from green to ovendry moisture content) | | |
	RADIAL	TANGENTIAL	VOLUMETRIC
HARDWOODS			
Alder, red	4.4	7.3	12.6
Ash:			
Black	5.0	7.8	15.2
Blue	3.9	6.5	11.7
Green......................	4.6	7.1	12.5
Oregon.....................	4.1	8.1	13.2
Pumpkin	3.7	6.3	12.0
White	4.9	7.8	13.3
Aspen:			
Bigtooth....................	3.3	7.9	11.8
Quaking....................	3.5	6.7	11.5
Basswood,			
American...................	6.6	9.3	15.8
Beech, American	5.5	11.9	17.2
Birch:			
Alaska paper	6.5	9.9	16.7
Gray	5.2	...	14.7
Paper	6.3	8.6	16.2
River	4.7	9.2	13.5
Sweet	6.5	9.0	15.6
Yellow	7.3	9.5	16.8
Buckeye, yellow	3.6	8.1	12.5
Butternut....................	3.4	6.4	10.6
Cherry, black	3.7	7.1	11.5
Cottonwood:			
Balsam poplar	3.0	7.1	10.5
Black	3.6	8.6	12.4
Eastern	3.9	9.2	13.9
Elm:			
American...................	4.2	7.2	14.6
Cedar	4.7	10.2	15.4
Rock	4.8	8.1	14.9
Slippery...................	4.9	8.9	13.8
Winged	5.3	11.6	17.7
Hackberry....................	4.8	8.9	13.8
Hickory, Pecan	4.9	8.9	13.6
Hickory, True:			
Mockernut	7.7	11.0	17.8
Pignut	7.2	11.5	17.9
Shagbark	7.0	10.5	16.7
Shellbark..................	7.6	12.6	19.2
Holly, American	4.8	9.9	16.9
Honeylocust.................	4.2	6.6	10.8
Locust, black	4.6	7.2	10.2
Madrone, Pacific	5.6	12.4	18.1
Magnolia:			
Cucumbertree	5.2	8.8	13.6
Southern	5.4	6.6	12.3
Sweetbay	4.7	8.3	12.9
Maple:			
Bigleaf	3.7	7.1	11.6
Black	4.8	9.3	14.0
Red.......................	4.0	8.2	12.6
Silver	3.0	7.2	12.0

SPECIES	RADIAL	TANGENTIAL	VOLUMETRIC
Striped	3.2	8.6	12.3
Sugar	4.8	9.9	14.7
Oak, red:			
Black	4.4	11.1	15.1
Laurel	4.0	9.9	19.0
Northern red	4.0	8.6	13.7
Pin	4.3	9.5	14.5
Scarlet	4.4	10.8	14.7
Southern red	4.7	11.3	16.1
Water	4.4	9.8	16.1
Willow	5.0	9.6	18.9
Oak, white:			
Bur	4.4	8.8	12.7
Chestnut	5.3	10.8	16.4
Live	6.6	9.5	14.7
Overcup	5.3	12.7	16.0
Post......................	5.4	9.8	16.2
Swamp			
chestnut..................	5.2	10.8	16.4
White	5.6	10.5	16.3
Persimmon,			
common...................	7.9	11.2	19.1
Sassafras	4.0	6.2	10.3
Sweetgum..................	5.3	10.2	15.8
Sycamore,			
American..................	5.0	8.4	14.1
Tanoak.....................	4.9	11.7	17.3
Tupelo:			
Black	5.1	8.7	14.4
Water	4.2	7.6	12.5
Walnut, black................	5.5	7.8	12.8
Willow, black................	3.3	8.7	13.9
Yellow-poplar	4.6	8.2	12.7

SOFTWOODS

Baldcypress	3.8	6.2	10.5
Cedar:			
Alaska-.....................	2.8	6.0	9.2
Atlantic white-..............	2.9	5.4	8.8
Eastern			
redcedar	3.1	4.7	7.8
Incense-	3.3	5.2	7.7
Northern			
white-...................	2.2	4.9	7.2
Port-Orford-	4.6	6.9	10.1
Western			
redcedar	2.4	5.0	6.8
Douglas-fir:			
Coast	4.8	7.6	12.4
Interior north..............	3.8	6.9	10.7
Interior west	4.8	7.5	11.8
Fir:			
Balsam....................	2.9	6.9	11.2
California red	4.5	7.9	11.4
Grand.....................	3.4	7.5	11.0
Noble	4.3	8.3	12.4
Pacific silver	4.4	9.2	13.0
Subalpine	2.6	7.4	9.4
White	3.3	7.0	9.8
Hemlock:			
Eastern	3.0	6.8	9.7
Mountain	4.4	7.1	11.1
Western...................	4.2	7.8	12.4
Larch, western	4.5	9.1	14.0

(Continued)

SHRINKAGE VALUES BY PERCENT (continued)

SPECIES	RADIAL	TANGENTIAL	VOLUMETRIC
Pine:			
Eastern white	2.1	6.1	8.2
Jack	3.7	6.6	10.3
Loblolly	4.8	7.4	12.3
Lodgepole	4.3	6.7	11.1
Longleaf	5.1	7.5	12.2
Pitch	4.0	7.1	10.9
Pond	5.1	7.1	11.2
Ponderosa	3.9	6.2	9.7
Red	3.8	7.2	11.3
Shortleaf	4.6	7.7	12.3
Slash	5.4	7.6	12.1
Sugar	2.9	5.6	7.9
Virginia	4.2	7.2	11.9
Western white	4.1	7.4	11.8
Redwood:			
Old-growth	2.6	4.4	6.8
Young-growth	2.2	4.9	7.0
Spruce:			
Black	4.1	6.8	11.3
Engelmann	3.8	7.1	11.0
Red	3.8	7.8	11.8
Sitka	4.3	7.5	11.5
Tamarack	3.7	7.4	13.6

Planning for dimension changes. This is essential for certain types of furniture. For example, if you are making a desk top from a wide piece of solid wood, or an edge-glued panel, and you glue edge bands at right

In edge gluing, moisture from glue soaks into the wood next to glue line, shown in the left-hand drawing. Soaked-in moisture from glue causes the wood to swell, shown in exaggeration at right.

If the wood is planed or level-sanded while the portion next to the glue line is still moisture-swollen, it remains flat only unit moisture leaves it. As shown at right, when moisture leaves the swollen area next to the glue line, the wood shrinks and leaves hollows where it was planed while still moisture-swollen.

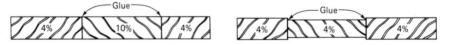

Edge-glued boards should not be planed while they have unequal moisture contents, as shown on the ends of the boards at left. Otherwise they will shrink unequally as they dry, as shown at right. Best bet is to use boards that have equal moisture contents before gluing.

angles across the ends of the panel, all's well so long as the moisture content of the pieces remains constant. But if the moisture content drops, as from humid summer to winter steam heat, the large panel will shrink considerably across the grain. Yet because the grain of the edge bands is lengthwise, the edge bands will shrink very little. The result is likely to be a split in the large panel. The same kind of problem can occur with a solid wood-ribbon top table banded and glued at right angles across the ends. To avoid splits, fasten the right-angled bands only at or near their midpoints, using several dowels or a tongue and groove to keep them aligned. The ends of the bands may not always be flush with the large panel edges, but it's seldom noticeable and helps you avoid splits.

To hold an edge-glued, ribboned tabletop flat and attach the legs with minimum chance of shrinkage splits, you can use slotted bolt-on cross bands. The slots allow ample movement for shrinkage, but nuts or bolts should not be overtightened.

Tips: Quarter-sawed wood shrinks less, so it should be favored over

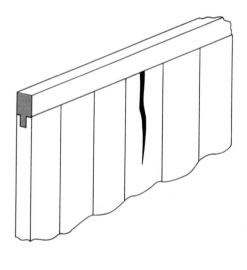

EDGE BANDS

If an edge band is glued to a wide panel with grain running at right angles, like this, splits are likely when humidity changes. The cause is the large difference in rates of shrinkage between grain and cross-grain pieces.

Differences in moisture contents at the time of gluing or differences in species can cause one member of an assembly to swell or shrink more than the other, as shown. Here a dowel is represented by dotted lines. *Preventive measure:* Use seasoned wood.

(Continued)

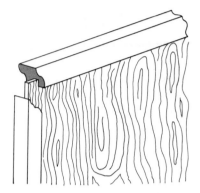

Where different rates of shrinkage can't be entirely avoided, overhangs or other design features can be used, as shown, to avoid obvious changes.

When a large panel can be housed in a grooved frame, like this, glue can be left out of grooves, and leeway can be allowed for unequal shrinking or swelling.

plain-sawed for the types of construction mentioned above. If you're using two species of wood in ribbon top work, favor species with fairly similar shrinkage. Where glue blocks are to be used at right angles to the grain of a corner joint, use a series of short blocks rather than one long one. The natural "give" of the wood over a short length can usually take care of shrinkage differences that might cause separation of a longer block.

This edge-glued ribbon-top table is made of white pine and stained cypress. The glue is resorcinol. End bands are tongue-and-groove fitted to ends of the ribbon top, bonded only at the midpoint. After more than 25 years, the table had developed no splits.

Checking for planer damage. On all wood to be glued, check the surface for evidence of a dull planer blade. This usually shows up as glazed areas. The glaze not only seals the surface, preventing a good glue bite, but results in flattened and less absorbent subsurface cells (from blade pounding). This is likely to cause a poor glue bond. Replaning or thorough sanding by machine or hand can solve the problem. If you sand, be sure you keep the corners square and sharp.

Glue joints. Wherever possible, glue joints should be planned to provide adequate gluing surface as well as direct wood support if sustained loads are involved, especially where thermoplastic glues are used, as in the case of typical white glues. The wood support is advisable with any type of glue. Of course, you have a wide choice of joint types that can provide that support.

Furniture frame design may be modified to provide the necessary resistance to normal stress. Rungs or stretchers are useful in this respect, as are flared rails at corner-leg joints and at other points.

Doweled joints. These are often limited in length by the joint dimensions. The joint strength increases with dowel length within a length-to-diameter ratio usually considered best at around 8 or 9 to 1, with some variation, because gluing area increases considerably with diameter. Typically, the maximum holding power of a ¼-inch dowel would be reached at around a 2-inch length. For a ⅜-inch dowel, this maximum would be 3 inches, and for a ½-inch dowel, it would be around 4 inches. Still, joints of adequate strength can be made with shorter dowels. Commonly used dowels include the plain types available at most hardware stores, the fluted type, and the spiral-cut type. The last two are designed to prevent the piston effect that often occurs when the dowel is pushed into the dowel hole. This is caused by the glue acting as a lubricant and sealant that traps air in the bottom of the hole. The resulting compression tends to push the dowel back out the hole as soon as you release your hand pressure. The fluted and spiral-cut dowels provide air-release channels. The same effect can be obtained by sawing a few lengthwise grooves in a plain dowel to release trapped air and glue. Chamfering of the dowel ends to eliminate the sharp corner also helps reduce the chance that the dowel will snag on roughness inside the hole.

Plan your dowel joint assemblies so you can coat both the dowel and the inside of the dowel hole with glue. In tight spots a bent-to-fit tobacco-pipe cleaner is often a good coating tool.

Try your joints dry, especially if you're new to glue work. Normally, joints such as mortise-and-tenon types should fit just snugly enough to require gentle hand pressure to make them fit.

CONSIDERING JOINT LOAD

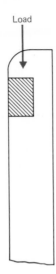

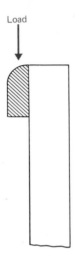

The best glued joints are supported by the wood itself. Glue simply holds parts in place. During gluing, joints should fit so as to allow parts to slide together without scraping off glue. Yet joint fit should not be loose. This will possibly require some hand fitting.

A weak joint, this one should not be used with white glue or with other glues that can creep under heat, humidity, and steady load. Here, glue takes the entire load. Thermosetting glues such as resorcinol can be used. Still, wood-supported joints are better.

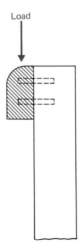

Here dowels, shown by dotted lines, take the load while glue holds the joint.

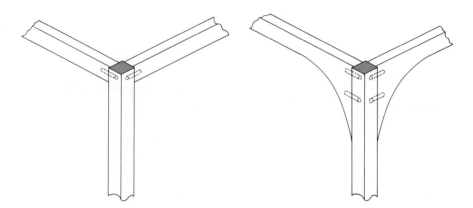

The joint at left is sometimes used in chair or table construction. Because of small cross-sectional size of parts, only one dowel is used. This is not a strong joint. The right-hand drawing shows how a side rail can be flared at the ends to increase leverage of the glue joint, thereby adding strength. This flaring also improves the appearance of the piece.

In planning glued joints for strength, design them to provide as much face grain as possible—such as the surface or edge of a board. In the case of dowel joints, which provide a fair amount of this kind of grain, try to have ample dowel length. Also, where feasible, design extra width into your dowel joints so more than one dowel can be used.

Very important in dowel work is effective spreading of the glue on the dowel and inside the dowel hole. If, for handling reasons, you can't completely coat the dowel, favor the coating inside the hole. Usually, however, it's easy to coat the dowel with a small brush. For the inside of the dowel hole, you can use a tobacco pipe cleaner or a brush of the type used to clean pump tubes of coffee percolators. You can buy the brushes at most hardware stores or from the houseware departments of most department stores.

Important before you start the doweling work, try the drill bit you'll be using on some scrap wood. Then try a section of your dowel in the hole. The dowel should slip in easily to allow space for the glue around it in the final job. If the dowel has to be driven in, the fit is so tight that it would wipe away the glue. If the dry fit is okay, coat the dowel and the hole. If the dowel bobs up like a piston after being pushed into the hole, trapped air is the culprit. To release air, either make a lengthwise saw cut about ⅛ inch deep from end to end of each dowel, or buy spiral-fluted or straight fluted dowels if you can find a nearby source. If you can't you can order them from woodworking supply houses such as Al-

USING A DOWELING JIG

The Dowel-it Model 1000 doweling jig allows precise, automatic centering of holes in 5 diameters on wood up to 2 inches wide. The drill guide itself is hardened to resist drill-bit wear. A Model 2000 is available with three removable screw-in bushings that allow drilling of 3 additional hole diameters in between sizes shown on the Model 1000, above. The units are available through retailers or the Dowel-it Company, PO Box 147, Hastings, MI 49058. Steps for use of the doweling jig are as follows.

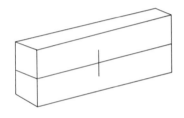

1. Mark mating wood pieces on one side, showing the axis for the drill hole.

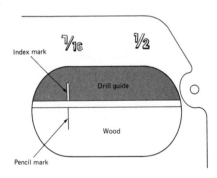

Index mark

Drill guide

Wood

Pencil mark

2. Clamp the Dowel-it over *one* piece of wood so your pencil mark lines up with a machined index mark visible on the drill guide itself when you look through the appropriate oval window.

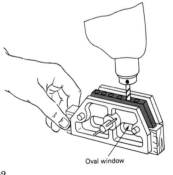

Oval window

3. Drill the dowel hole to the desired depth. Note: You can ensure correct depth by first marking the drill bit with a ring of masking tape. Then repeat steps 2 and 3 to drill the hole on the mating piece of wood.

bert Constantine and Son, 2050 Eastchester Rd., Bronx, NY 10461. They are usually stocked in batches of ready-to-use lengths, commonly from 1½ to 3 inches, with diameter up to about ⅜ inch.

If your bit makes holes too tight for the dowels you're using, try running the bit up and down in the holes a few times. This often shaves the hole just enough to do the trick. You don't want a loose-fitting hole that results in a weak glue bond. For an accurate fit in most doweling work, you'll need a doweling jig. These come with guides for drill bits of various diameters.

Where parts of your project are mitered lengthwise (with miter running parallel to the grain) you can usually make a glued miter joint as

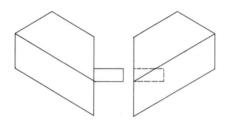

A plain miter joint may be glued and reinforced with a dowel (as shown), or else a screw or nail.

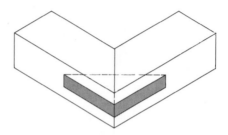

A miter joint with slip feather is easy to make. After making the miter cuts, run the mitered ends through a table saw to cut the slots for the slip feather. Then coat slip feather with glue and push into place.

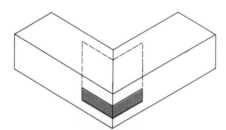

If the miter is splined, cuts for the spline must be made before the joint is glued. Cut and trim the spline dry. Then glue it and sand it flush.

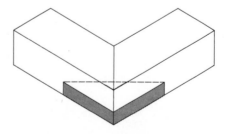

Gusset-type reinforcement may also be used on the back of a miter. A back saw and chisel are the only tools you will need. Or you can use a router.

strong as the wood or stronger. When the miter is at the ends of pieces, however, as at the corners of a picture frame, the meeting surfaces are essentially end grain and the joint has considerably less strength. Clamping the joint squeezes out excess glue. But many slow setting glues continue to soak into the porous end grain until there isn't enough glue between the surfaces. This results in a starved, low-strength joint. One remedy sometimes used in commercial work is that of using a high-solids fast-setting glue for miters. Another method that's easier in the home shop is a sizing coat of the glue you're using. Allow it to dry briefly, then recoat the surfaces and assemble the joint.

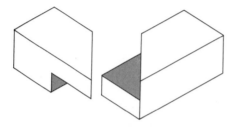

A mitered half lap can be made on a table saw. This is a strong joint because it has face-to-face gluing surfaces. It looks like a simple miter from the front and like a corner lap joint from the back.

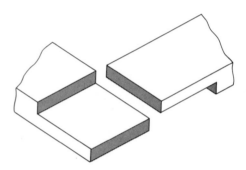

End lap joints of this and other types are strong because glue bonds large face-to-face areas, rather than merely bonding end grain.

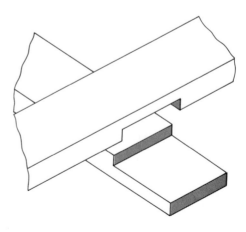

The middle lap also allows strong face-to-face gluing.

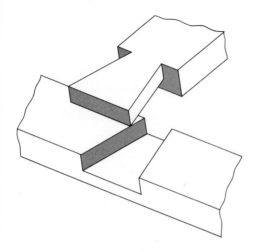

Though time-consuming to make, the dovetail half lap has ample face-to-face gluing area, good strength, and presents a neat appearance.

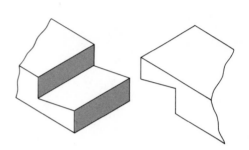

The beveled lap has good strength, although it is rarely needed.

Providing face-to-face gluing area offers the strongest joints. This can be done by various methods, some of which are shown in accompanying drawings. One of the easiest (suited to light loads) is the slip feather. The slot for it can be cut by hand or power saw after the miter is made. The glue-coated slip feather may then be inserted once or twice to spread the glue, and finally to remain as a permanent part of the joint. A relatively thin slip feather (saw kerf thickness) is adequate for projects such as picture frames, including old frames that come apart at the corners.

DADO JOINTS

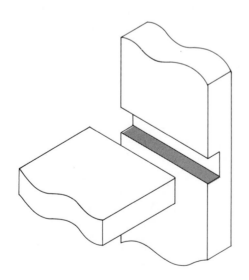

The plain dado is often used for shelves. This is a weak glue joint because all surface grain meets end grain. Screws into shelf ends are advisable here.

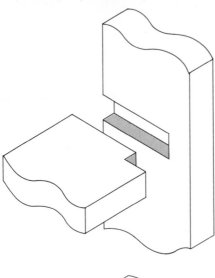

A stopped dado isn't visible from the stopped end, so it looks better in bookcases. Yet it has the same deficiencies as the plain dado.

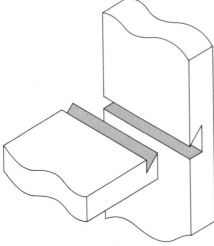

A dovetail dado won't pull out because of the wood lock formed by the cuts. But the holding power of its glue is about the same as for other dadoes shown.

A single dovetail joint is held together by the lock formed by the wood cuts as well as by the glue. Glue merely holds parts in place. The wood itself supports the load.

A multiple-tenon finger joint provides surface grain area for bonding. This is a popular joint for dresser drawers.

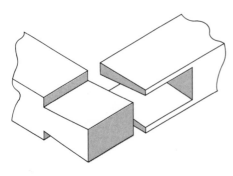

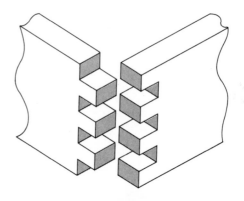

12 | General Repairs

TODAY'S REPAIR GLUING may involve almost anything from a rocking chair to a pair of earrings—a china ornament to a car body. Years ago, many of these same jobs could not be done with glues. Some required brazing or soldering, and others called for hours or days of work rather than the minutes often possible with modern glues. Fortunately, there's now a glue, adhesive, or cement for just about every repair. Selecting the right one for the job is the first step. In some cases, as in the car-body work described next, just knowing how to do the work can save you money—even before you do it.

CAR BODY HOLES. These can hold down the selling price of a used car and may actually make it a bargain if you know how to patch it. Many auto-supply outlets sell small kits (including the adhesive) for small jobs. Simply follow the manufacturer's instructions. If you're facing a major job (big holes or lots of small ones) you may save some money by using the materials and instructions that follow. (There are other methods, but I've found that those below can conceal holes on cars exposed year-round to the weather for periods of 5 years.)

Adhesive basics. You can use the same polyester resin that is used to fiberglass boats. And you can use fiberglass window screening as a base material over the car body rust hole if it's in a flat area of the car body. When considerable shaping is necessary, as for a compound curve, aluminum screening serves better because it can be bent to shape and will usually retain the shape. Do the shaping by eye, and take your time. Bond either type of screening on with polyester adhesive, as I'll describe shortly. Small holes, up to 1½ inches in diameter may not require screening for a base, which saves time. Sometimes these minor jobs can be completed in about an hour. The final covering is fiberglass cloth applied over the screening (or over small holes without screening) and bonded with polyester resin. Now for the details.

• *The first step.* Clean and disk-sand the painted area around the hole. (Use a detergent if the area's greasy.) If the paint is in good firm condition, you won't have to sand to the bare metal except in an area about an inch wide around the hole. Use a medium grit aluminum oxide abrasive

107

disk. Don't bear down too hard on the disk (just enough to flex it slightly). Otherwise, you may find that the metal under the paint is so badly pitted from the other side that pieces shear off. Simply provide a clean, moderately roughened surface to which you can resin-bond your screen and fiberglass. You are actually rebuilding a part of the car body with the fiberglass and screen, and if you do it carefully it will probably be stronger than the original.

• *The second step.* After cleaning and sanding, prepare and bond the patch. If you're using screening as a base, drill a few holes around the rust hole perimeter for temporary sheet metal screws that will hold the screening in place while the first coat of resin hardens. The screening should be held firmly against the metal with about a 1½-inch overlap around the hole.

With the screen patch in place, apply the resin (by brush) mixed according to the manufacturer's instructions. But *do not* apply it to the screws or to the areas immediately surrounding them. Be sure to work the resin through the screen so it will bond the screen to the metal around the hole. When the resin is hard enough to hold the screen (typically about half an hour) remove the sheet metal screws. They're intended only to hold the screen until the resin hardens. If there are any projecting snags of screen, wait for the resin to harden a little more. Then sand off the snags carefully. After this, apply another coat of resin extending about 2 inches beyond the screen, and smooth a cut-to-fit fiberglass patch onto the resin. This covers the screw holes and provides the surface that will later be painted. When this fiberglass bonding coat hardens, apply another generous coat to the fiberglass, extending it a little beyond the patch. If, when hard, this coat still shows the fabric pattern, you can squeegee on a smoothing coat of auto-body putty, and later sand it smooth. If you're not in a hurry an extra coat of resin prior to the putty is a good idea. This may eliminate the need for the putty. (If the rust hole area is very large you can use two layers of fiberglass instead of one.)

In any event, the perimeter of the patch must be feathered out with the disk sander to blend smoothly with the surrounding metal body surface. Use a medium-grit disk for the bulk of the feathering and follow with a fine disk to remove scratch marks. If any remain fill them with body putty. After final smooth sanding, the surface is ready for painting, subject to a waiting period sometimes specified by the resin manufacturer.

• *Tips and cautions.* Before doing any drilling in the auto body, make sure the drill bit can't suddenly push through and hit the gas tank, fuel line, wiring, or other vulnerable part. Also, if you smell gasoline, don't tackle the body work until you've located the source of the fumes and

This auto-body hole is being patched with screen, overlain with fiberglass cloth. The chrome molding holds the bottom of the patch. The end of the screen is folded around a corner of the door opening.

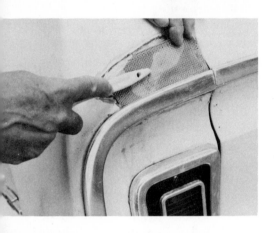

To patch the hole above this tail light, the lower edge of the screen was slipped under the molding. Resin was applied through the screen and pressed in firm contact with brush handle.

The entire base of this fender skirt is built up with screen. Packaging tape, used to hold the upper edge of screen in place, is removed after the resin hardens. Fiberglass cloth is applied over the screen.

The final step is in smoothing and feathering the edges of the fiberglass to blend with metal of the car body. Use medium grit abrasive disks first, proceeding to fine grit.

109

eliminated it. Sparks from tools or sanding can cause a vapor flashback to the fuel tank with disastrous results.

You can buy your resin and fiberglass from marine or auto supply dealers, boatyards, or mail-order houses. The fiberglass is available in various weights, by the square yard or in tape form. The square yard widths generally run from 38 to 60 inches, though you're not likely to need such sizes for auto-body repair. Tape widths commonly run from 1½ to 12 inches. For body work (selected according to the job) you'll probably need widths from 4 inches up. The weight for this type of work is usually in the 9- to 10-ounce-per-square-yard range.

The resin is available in pints, quarts, and gallons. For auto-body work, you'll usually be working with pints and quarts. Mix only as much as you can use well within the hardening time specified on the container, typically 15 or 20 minutes, depending on temperature. (Tin cans make good mixing vessels.) The catalyst hardener is usually packaged in tubes marked for use with fractional portions of the resin (such as half a tube to half a can of resin). If you change brands between jobs or between portions of the same job, follow the instructions on the brand you're using. There can be quite a difference between brands. Check especially on hardening times and temperatures. Most brands begin to harden in about 30 minutes at 70°F., and are hard enough for sanding and drilling in around 3 hours, though some work faster.

GLUE REPAIRS ON WOODWORK. Loose furniture joints, probably the commonest subjects for glue repair, are much easier to correct now than before the development of synthetic resin glues. Today, for example, it is seldom necessary to dismantle a loose joint in order to make it tight, often a job that required extensive disassembly of the piece.

Instead of disassembly, you can now usually use a modern gap-filler adhesive to tighten a loose joint, and in most instances, hold it as well or better than the old adhesive. The modern adhesive should be a stick-to-anything type such as epoxy or acrylic. It can then bond to the old separated glue-coated surfaces and also fill the gap between them.

In the typical instance of a chair rung that has loosened in the chair leg, you simply tip the chair and prop it in a position that enables you to run the gap-filler adhesive into the gap in the loose joint. After it's in, wriggle the joint parts (leg or rung) to be sure the glue reaches all parts, and add more glue if necessary. A small piece of aluminum flashing can often be shaped to serve as a miniature funnel. After that you need only wait for the glue to harden for the time specified by the manufacturer, and the job's done. You are literally forming a plastic bushing between

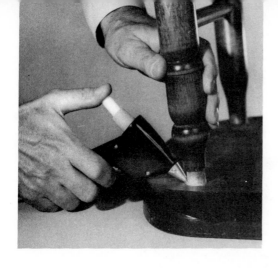

If legs or rungs are loose in holes, pull them partway out, like this, applying a quick-setting glue and pushing them back into place.

the loose parts. You can use an acrylic such as P.A.C. (Tridox Laboratories, 212 N. 21st St., Philadelphia, PA 19103) or quick setting epoxies such as Fibre Glass-Evercoat's Epoxy Paste Glue (Fibre Glass-Evercoat Co. Inc., Cincinnati, OH 45242), or the Clear Epoxy Kits (Loctite Corporation, Newington, CT 06111). These kits are packaged in pull-top cans containing 10 flat 2-inch diameter individual-job kits, each with enough of the twin components for an average household repair job. Just peel off the plastic kit cover, mix the components, and apply. The remaining kits stay sealed until used. The glue's setting time: 5 minutes.

The glue injector. This can be used for many repairs where regluing is essential without disassembly. It's best to use only glues that can be removed from the tool with water or an effective solvent. To prepare the joint, simply drill a hole in the joint to accept the injector's nozzle. Glue can then be forced in by the piston action of the injector, forcing glue to the internal parts of the joint. Several injections may be made, but be sure to clean the injector before the glue begins to set in it. Typical glue injectors, such as those sold by Albert Constantine and Son, 2050 Eastchester Rd., Bronx, NY 10461, are made of metal, and can be cleaned with painter's lacquer thinner containing toluene or toluol. A Sherwin-Williams product of this type is called SW-Toloul R2K1.

The glue injector is not limited to furniture work, but can be used for a wide variety of applications where you need to force an adhesive into loose-fitting parts to lock them against unwanted motion, as in squeaky floors. Just rock your weight on the flooring board where you can produce a squeak. Then bore a hole (of injector nozzle size) through the flooring board, and inject the adhesive. A gap filler such as casein is good. Allow the adhesive to set, and test for the squeak again. One treatment usually cures it. Epoxy is also good if you have the toluol at hand to clean your glue injector. The same types of adhesives can also be used

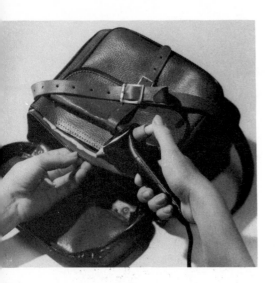

A gun glue is flexible enough to bond flexible parts such as this purse zipper. Allow time after application to bring the parts firmly together.

for bric-a-brac repairs. You can use the usual wood glues for wooden objects, stick-to-anything types such as epoxy or acrylic for other materials. If you're working on plastic, try the adhesive on a small, inconspicuous area to be sure the solvent used in it doesn't damage the plastic. Where a clean break in a plastic object is to be repaired, and you know the type of plastic that's broken, buy an adhesive that's made for that type of plastic, styrene, etc. You'll usually find the types of plastic for which the adhesive is suited printed on the tube. Such adhesives can often fuse the parts together almost as if the break had not occurred. If you don't know the type of plastic to be repaired, try several of the plastic adhesives on inconspicuous parts of the broken piece, allow each one to remain for a minute or so, and wipe it off. Adhesives that etch the plastic surface will usually bond to it.

CHINA, CROCKERY, AND GLASS. These can also now be repaired with adhesives that do the job better than was possible a few years ago. But even though the repair can often be made almost invisible, don't expect it to have the full strength and durability of the unbroken piece. A cream pitcher with a broken handle bonded back in place, for example, may be strong and waterproof but not necessarily dishwasher-proof.

The first step is to select the glue for the job. If the object being repaired is purely decorative, and broken in a number of pieces, you might do well with a vinyl acetate such as Miracle Adhesives Corporation's Sheer Magic (Bellmore, Long Island, NY 11710). In assembling a number of broken parts you apply the clear adhesive to the broken edges and allow it 2 or 3 minutes to become tacky. Small broken fragments put together at this stage will usually be held in place simply by the tackiness of the adhesive—a great help in repairing badly shattered china ornaments. The strength of the bond increases for hours, sometimes days, depending on the job. Allow at least an overnight period. You can remove excess adhesive with lacquer thinner. *Very important* in china and crockery repairs: Save *all* the broken pieces—even those so small you have to sweep them up in a dust pan. They can be fitted in place during the repair job. A missing piece makes the repair obvious. In general, a clear adhesive simplifies repair work that must be inconspicuous because the adhesive does not conflict with varying color areas of the object. Press the parts together tightly to make the glue line as nearly invisible as possible.

If you want a waterproof job, favor an epoxy with an average setting time of 30 minutes or more. (The fast, 5-minute types are not usually recommended for jobs requiring high water resistance.) If you're in doubt about waterproofness, check with the manufacturer. In using adhesives of this type, it's necessary to provide a reliable means of holding the parts together because few such epoxies can be used for repair work relying on tack alone to hold the parts together. The holding devices depend on your ingenuity. Weights are the most widely used for irregular shapes often encountered in broken china and crockery. But remember you can use a large weight (such as a can of paint) for the base

Parts of this ornamental dog are held in position with wire solder while the adhesive hardens. Bend the solder before applying adhesive. The solder forms a cradle and positions the dog's limbs.

Use your ingenuity in assembling weights and supports when repairing china ornaments. Most adhesives suited to the job don't need clamping—just enough pressure to keep broken parts in firm contact. Here, the umbrella of a Hummel ornament is held in position by wrenches, a flat stone, and a plane, with cans as supports. Many types of adhesives are suitable, as covered in Chapter 2.

support, with another movable weight on top. Wire solder is also a versatile holding device. It's easily bent to grip the weight used as a base, and easily bent to grip almost any shaped object. When the glue has set, the solder can be removed and later reused for other projects. For repairs of heavier objects, you can use aluminum wire of the type often used for TV antenna bracing. It's almost as soft and easily shaped as the wire solder, but it's just enough stiffer to handle heavier work. In a pinch you can also use ordinary copper electrical wire after stripping the insulation.

LOOSE KNIFE HANDLES. These are frequently on the list of necessary repairs around the house, and they are often much simpler to repair than they might seem. Usually they involve a knife whose blade is anchored simply by a tang that fits into a hole in the handle.

A thoroughly clean gluing surface is the first essential. Once the tang (integral with the blade) is removed from the handle, cleaning isn't difficult. Clean the tang with household detergent, and make sure no grease deposits remain. Use detergent also to clean out the hole in the handle. Depending on the size of the hole, you can use either a pipe cleaner for internal scrubbing, or a percolator brush (made for cleaning the tube in a percolator).

After the tang and the handle hole are clean, allow them to dry thoroughly. You can hasten the handle drying by shaking the water out and mopping with cotton or a Q-tip. When the tang is completely dry, coat it and the inside of the hole with a waterproof epoxy, and support the handle (hole up) in a vise or drilled block. Use plenty of glue and wipe off any overflow. Prop the blade carefully in alignment with the handle. If you want a quick job you can use a waterproof acrylic such as P.A.C.

114

Index